T0313718

# Advances in Solid Oxide Fuel Cells VI

# Advances in Solid Oxide Fuel Cells VI

*A Collection of Papers Presented at the 34th International Conference on Advanced Ceramics and Composites January 24–29, 2010 Daytona Beach, Florida*

Edited by
Prabhakar Singh
Narottam P. Bansal

Volume Editors
Sanjay Mathur
Tatsuki Ohji

The American Ceramic Society

A John Wiley & Sons, Inc., Publication

Published by John Wiley & Sons, Inc., Hoboken, New Jersey.
Published simultaneously in Canada.

For general information on our other products and services or for technical support, please contact our Customer Care Department within the United States at (800) 762-2974, outside the United States at (317) 572-3993 or fax (317) 572-4002.

Wiley also publishes its books in a variety of electronic formats. Some content that appears in print may not be available in electronic format. For information about Wiley products, visit our web site at www.wiley.com.

*Library of Congress Cataloging-in-Publication Data is available.*

ISBN 978-0-470-59469-8

Printed in the United States of America.

10 9 8 7 6 5 4 3 2 1

# Contents

# Preface

The Seventh International Symposium on Solid Oxide Fuel Cells (SOFC): Materials, Science, and Technology was held during the 34th International Conference and Exposition on Advanced Ceramics and Composites in Daytona Beach, FL, January 24 to 29, 2010. This symposium provided an international forum for scientists, engineers, and technologists to discuss and exchange state-of-the-art ideas, information, and technology on various aspects of solid oxide fuel cells. A total of 75 papers were presented in the form of oral and poster presentations, including ten invited lectures, indicating strong interest in the scientifically and technologically important field of solid oxide fuel cells. Authors from eleven countries (China, Denmark, Germany India, Italy, Japan, Russia, South Korea, Taiwan, UK and U.S.A.) participated. The speakers represented universities, industries, and government research laboratories.

These proceedings contain contributions on various aspects of solid oxide fuel cells that were discussed at the symposium. Fifteen papers describing the current status of solid oxide fuel cells technology are included in this volume.

The editors wish to extend their gratitude and appreciation to all the authors for their contributions and cooperation, to all the participants and session chairs for their time and efforts, and to all the reviewers for their useful comments and suggestions. We hope that this volume will serve as a valuable reference for the engineers, scientists, researchers and others interested in the materials, science and technology of solid oxide fuel cells.

PRABHAKAR SINGH
*University of Connecticut*

NAROTTAM P. BANSAL
*NASA Glenn Research Center*

# Introduction

This CESP issue represents papers that were submitted and approved for the proceedings of the 34th International Conference on Advanced Ceramics and Composites (ICACC), held January 24–29, 2010 in Daytona Beach, Florida. ICACC is the most prominent international meeting in the area of advanced structural, functional, and nanoscopic ceramics, composites, and other emerging ceramic materials and technologies. This prestigious conference has been organized by The American Ceramic Society's (ACerS) Engineering Ceramics Division (ECD) since 1977.

The conference was organized into the following symposia and focused sessions:

| | |
|---|---|
| Symposium 1 | Mechanical Behavior and Performance of Ceramics and Composites |
| Symposium 2 | Advanced Ceramic Coatings for Structural, Environmental, and Functional Applications |
| Symposium 3 | 7th International Symposium on Solid Oxide Fuel Cells (SOFC): Materials, Science, and Technology |
| Symposium 4 | Armor Ceramics |
| Symposium 5 | Next Generation Bioceramics |
| Symposium 6 | International Symposium on Ceramics for Electric Energy Generation, Storage, and Distribution |
| Symposium 7 | 4th International Symposium on Nanostructured Materials and Nanocomposites: Development and Applications |
| Symposium 8 | 4th International Symposium on Advanced Processing and Manufacturing Technologies (APMT) for Structural and Multifunctional Materials and Systems |
| Symposium 9 | Porous Ceramics: Novel Developments and Applications |
| Symposium 10 | Thermal Management Materials and Technologies |
| Symposium 11 | Advanced Sensor Technology, Developments and Applications |

Focused Session 1  Geopolymers and other Inorganic Polymers
Focused Session 2  Global Mineral Resources for Strategic and Emerging
                   Technologies
Focused Session 3  Computational Design, Modeling, Simulation and
                   Characterization of Ceramics and Composites
Focused Session 4  Nanolaminated Ternary Carbides and Nitrides (MAX Phases)

The conference proceedings are published into 9 issues of the 2010 Ceramic Engineering and Science Proceedings (CESP); Volume 31, Issues 2–10, 2010 as outlined below:

- Mechanical Properties and Performance of Engineering Ceramics and Composites V, CESP Volume 31, Issue 2 (includes papers from Symposium 1)
- Advanced Ceramic Coatings and Interfaces V, Volume 31, Issue 3 (includes papers from Symposium 2)
- Advances in Solid Oxide Fuel Cells VI, CESP Volume 31, Issue 4 (includes papers from Symposium 3)
- Advances in Ceramic Armor VI, CESP Volume 31, Issue 5 (includes papers from Symposium 4)
- Advances in Bioceramics and Porous Ceramics III, CESP Volume 31, Issue 6 (includes papers from Symposia 5 and 9)
- Nanostructured Materials and Nanotechnology IV, CESP Volume 31, Issue 7 (includes papers from Symposium 7)
- Advanced Processing and Manufacturing Technologies for Structural and Multifunctional Materials IV, CESP Volume 31, Issue 8 (includes papers from Symposium 8)
- Advanced Materials for Sustainable Developments, CESP Volume 31, Issue 9 (includes papers from Symposia 6, 10, and 11)
- Strategic Materials and Computational Design, CESP Volume 31, Issue 10 (includes papers from Focused Sessions 1, 3 and 4)

The organization of the Daytona Beach meeting and the publication of these proceedings were possible thanks to the professional staff of ACerS and the tireless dedication of many ECD members. We would especially like to express our sincere thanks to the symposia organizers, session chairs, presenters and conference attendees, for their efforts and enthusiastic participation in the vibrant and cutting-edge conference.

ACerS and the ECD invite you to attend the 35th International Conference on Advanced Ceramics and Composites (http://www.ceramics.org/icacc-11) January 23–28, 2011 in Daytona Beach, Florida.

Sanjay Mathur and Tatsuki Ohji, Volume Editors
July 2010

# SOLID OXIDE FUEL CELL (SOFC) BASED POWER SYSTEMS FOR MOBILE APPLICATIONS

R.M. Miller and T.L. Reitz
Air Force Research Laboratory (AFRL)
Wright-Patterson AFB, OH, USA

## ABSTRACT

As the largest user of energy within the DoD, the USAF consumes approximately 2.8 billion gallons of fuel annually in support of both domestic and deployed operations[1]. One solution being explored for reducing this energy need is to implement solid oxide fuel cell (SOFC)-based power systems. SOFCs can offer higher efficiencies when compared to conventional power generation approaches and greater compatibility with military logistic fuel than low temperature fuel cells. Efforts at AFRL are focused on increasing the power density of SOFC-based systems and on improving the operability of these systems on conventional battlefield fuels. Current SOFC-based activities are focused on developing a 2 kW compact, lightweight SOFC system for use in UAV prime power or vehicle APU applications. In addition, basic research efforts are also underway which seek to decrease the interfacial resistance and improve the operational flexibility of SOFCs. The objective of this paper is to present an overview of AFRL's activities which include SOFC system development and basic R&D activities focused an improving SOFC stack technology.

## INTRODUCTION

Solid oxide fuel cell (SOFC)-based systems offer an attractive alternative for internal combustion engines as field power generators, ground vehicle[2] and aircraft[3,4] auxiliary power units (APUs) and primary power units for small unmanned air vehicles (S-UAV). SOFC systems represent a compelling power system option due to their high efficiencies, fuel flexibility and low audible signature. Compared with other fuel cell approaches, the thermal environment and conductivity mechanism in SOFCs allow for a considerable improvement in fuel tolerance, providing a path forward for electrochemical logistic fuel operation.

Fuel cells are devices which electrochemically combine fuel and air to produce high quality electrical power. Because these systems do not generate power via combustion processes, they offer significantly lower specific fuel consumption rates relative to advanced heavy fuel engines or diesel power generators. Solid oxide fuel cells, in particular, offer the potential for fuel-flexible operation with the capability to operate off of hydrogen or reformate (containing hydrogen and carbon monoxide), supplied externally or produced internally via an onboard fuel reformer. Numerous fuel options exist including conventional or renewable hydrocarbon fuels such as standard gasoline or diesel, JP-8 (military diesel), ethanol, methanol, natural gas, propane, biodiesel, and Fisher-Tropsch synthesized fuels (e.g. dimethyl ether).

The Department of Energy (DoE) initiated the Solid State Energy Conversion Alliance (SECA)[5] program in 1999 to promote the advancement of SOFC power systems for a variety of energy needs. However, the emphasis of this alliance is on stationary power generation as opposed to mobile auxiliary power which is the primary need of the Department of Defense (DoD). Even so, the DoD can leverage the extensive efforts of the SECA program and related international efforts which have developed the base technology and manufacturing capabilities required to move forward. However, unlike the DoE goals of extremely long-life (<4%/1000hr) and low cost (<$400USD/kW)[6], the DoD has made investments to extend the operation of these systems to challenging military environments which include compact packaging as well as compatibility with fuels that may contain as much as 3000 ppm sulfur.

The objective of this paper is to present work performed at the Air Force Research Laboratory (AFRL) focused on the development of a SOFC-based power system for mobile applications. The

1

primary focus is on systems in the 1kW – 10kW range for UAV prime power or vehicle APU applications. In the coming section, a brief summary of the required goals and metrics needed to successfully produce a SOFC-based power system for mobile applications will be described. In addition, an overview of an ongoing cooperative development program between AFRL, the Army Tank Automotive Research, Development, and Engineering Center (TARDEC), and United Technologies Research Center (UTRC) will be reviewed. Lastly, efforts at AFRL focused on developing an improved SOFC stack for use in such a system are illustrated. These include exploring interfacial modification of SOFCs for increased performance and the exploration of alternate anode materials capable of operating in higher sulfur environments.

MOBILE SOFC SYSTEM DEVELOPMENT

The current AFRL system development program, jointly funded with Army TARDEC, is focused on the advancement of high power dense SOFC systems for use in UAV and silent watch APU applications. The ultimate objective of the program is to develop a power dense SOFC-based power system capable of operating on logistic fuels for mobile applications. The UTRC-led team has constructed a "packaged" bench top APU unit, which is shown in Figure 1. This unit is capable of producing 1.5 kW peak power while operating on S-8 (synthetic JP-8 produced using the Fischer-Tropsch process) fuel. S-8 was chosen as the initial fuel to remove issues pertaining to sulfur content in JP-8 and diesel fuel, allowing us to concentrate on issues related to increasing power density and reducing system size and footprint. Future iterations will look to expand upon the system design to incorporate desulfurization technologies or sulfur-tolerant stack technologies, such as those mentioned in the Sulfur-Tolerant Anode Materials section below.

The following sections will explore the motivation behind utilizing fuel cell-based power systems for mobile applications and give an overview of the current status of the AFRL mobile SOFC system development program.

Figure 1. 1.5 kW SOFC Bench Top System

System-Level Impact

When compared to traditional internal combustion-based approaches, fuel cell power systems offer higher efficiencies in the 1kW – 10 kW range of mobile power generators, but system power density is generally lower. In the case of stationary power systems, this trade is more acceptable due to the alternative advantages of fuel cell-based power systems such as quiet operation and reduced fuel consumption. However, for mobile applications such as vehicle mounted APUs, aircraft APUs and S-UAV prime power systems, the increased system weight can offset some of the advantages gained by the higher system efficiency. This is due to the fact that a larger APU or prime power system can increase the overall vehicle weight leading to increased fuel consumption or require the vehicle to carry a smaller payload. This is particularly true for aircrafts or UAVs where weight is a crucial factor in platform fuel consumption.

One method to compare the performance of an internal combustion (IC)-based system versus a fuel cell (FC)-based system is to look at the power system in terms of total mass, which includes the dry system mass and the fuel mass required to achieve a given level of endurance between refueling. Figure 2 compares the performance of a baseline IC system to a number of projected FC systems. The baseline IC system is sized assuming a power density of 1000 W/kg and a thermal efficiency of 15% (corresponding to a specific fuel consumption of ~ 1 lb/hp-hr for JP-8). The baseline FC system is sized to match the performance goals stated by United Technologies Research Center for their next generation mobile power system[7], which is a power density of > 100 W/kg with a thermal efficiency of ~ 30%. For a 2 kW system, the IC system has a dry weight of 2 kg while the FC system is an order of magnitude greater at 20 kg. Even with this difference, the increased efficiency of the FC system shows a cross over point at 33 hrs of operation. This crossover point represents the endurance at which the total mass (dry weight and fuel) of the IC system and FC system are equal. Therefore, from a purely mass-based point of view, for missions greater than 33 hrs between refueling the FC system is advantageous.

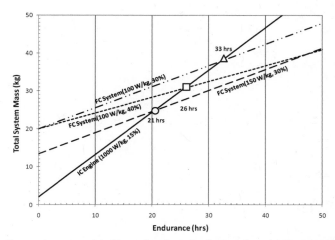

Figure 2. Total mass (system and fuel) as a function of endurance for a 2 kW representative internal combustion (IC)-based propulsion system and 2 kW representative fuel cell (FC)-based prime power systems.

Exploring the effect of increasing either the thermal efficiency (from 30% to 40%) or power density (from 100 W/kg to 150 W/kg) on the FC system, it can be seen that increasing the efficiency while holding the power density constant decreases the crossover point to 26 hrs while increasing the power density with a constant efficiency decreases the crossover point to 21 hrs of operation. This illustrates that at this size level, increases in power density provide the most impact on the total mass of the system. This is further illustrated in Figure 3, which shows plots of endurance as a function of power density or thermal efficiency, for a 2 kW system when the total mass is fixed at 25 kg. These plots illustrate that, for a given fixed overall mass, increases in power density from 100 W/kg to the range of 400 – 500 W/kg result in the largest impact to endurance, after which the gains start to level off. In addition, it can be seen that as the power density of the system increases the impact that gains in thermal efficiency have on the system endurance are greater. This shows that initial FC system development for mobile applications should primarily focus on increasing power density while maintaining efficiency. Once power density levels >200 W/kg are achieved, then thermal efficiency increases will start to show greater system level payoffs.

System Development Status

UTRC has successfully built and tested a 1.5 kW bench top system capable of operation on S-8 fuel[8,9]. The schematic in Figure 4 illustrates the design of the system. It is a single pass configuration utilizing a catalytic partial oxidation (CPOx) reformer to break down the fuel into a hydrogen-rich stream to be feed into the SOFC anode. Excess hydrogen from the anode stream is combusted in the burner to provide heat to the system. The burner exhaust is routed through a heat exchanger to provide heat to the incoming air stream which is routed to the SOFC cathode and CPOx inlet. The hot section components, which include: the CPOx reformer, SOFC stack, burner, and high-temperature heat exchanger, are enclosed in an insulated "hot box" (the larger section to the left in Figure 1) to maintain the required stack operating temperature (~ 800°C). This allows for a compact system design, but requires care to be taken to maintain the required thermal balance between the components. The cold section components are housed in a separate enclosure (the smaller section to the right in Figure 1) and include the fuel pump, air blower, valves and control electronics. A simple single pass system design with a CPOx reformer was chosen to make the system less complex and more compact. Other alternative designs include utilizing an autothermal reformer (ATR) with an anode recycle stream, which would boost the system efficiency but at the expense of added complexity and possibly added weight. As noted above, the biggest initial system level gains for fuel cell-based power systems is in increasing the system power density, therefore any design changes focused on increasing the system efficiency would have to maintain the current system level power density.

This initial bench top system was able to achieve a 26% system efficiency on S-8 at a power density of >50 W/kg. The efficiency was lower than the initial design target of 30%, but illustrates a key point for a thermally integrated system, such as this. One component does not operate independently of the others. The lower than expected efficiency was due to a requirement to run the reformer at an air to fuel ratio which was outside of the original design specifications. This led to an inability to maintain the stack temperature at the level required for higher efficiency operation due to the coupled interactions of the components and gas streams. The ability to maintain the proper thermal balance between components is a major challenge for the design of a compact SOFC system and will continue to play a significant role in future design efforts.

The current focus of the program is to further decrease the weight and volume of the system in order to produce a system at >100 W/kg capable of producing ~2 kW net power while operating on S-8. This effort is underway, with a goal of developing a system for flight demonstration during the summer of 2010. One key aspect for increasing the system power density is to further push the SOFC stack technology. The bench top system utilizes an SOFC stack developed by Topsoe Fuel Cell based on their planar SOFC stack design for mobile applications[10]. The bench top system stack operates at

~200 W/kg, which corresponds to ~30% of the system weight for a 50 W/kg system. To reach the weight targets for the flight system at 100 W/kg requires a stack that can produce ~300 W/kg corresponding to a stack that is almost 40% of the total system weight. This illustrates how critical power dense SOFC technology is to developing future mobile SOFC systems. To reach the current goal of a system-level power density >200 W/kg, it is estimated that an SOFC stack with a power density of >500W/kg is required assuming that the stack remains ~40% of the overall system weight.

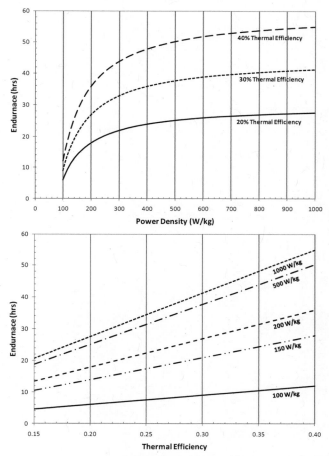

Figure 3. Plots illustrating the relationship between endurance and power density (top), thermal efficiency (bottom) for a 2 kW system with a fixed total mass (system and fuel) of 25 kg.

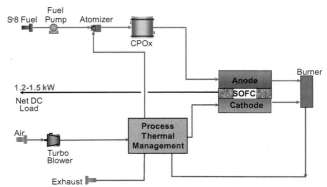

Figure 4. Power Dense SOFC System Design

Advances in stack power density alone will likely not be sufficient to achieve the long term objective of >200W/kg operating on JP-8. To achieve JP-8 operation it will be necessary to mitigate SOFC stack degradation associated with the high sulfur content in JP-8, which can reach 3000 ppm in the liquid phase which corresponds to ~300 ppm in the gas phase (primarily as $H_2S$) after the fuel reformation process. Several approaches are under development which show promise for achieving sufficient sulfur tolerance through improvements in stack material sets, cell structural changes, or stack operating modes[11,12,13]. These approaches, when combined with advanced reforming technologies, have the promise to enable JP-8 operation without the need for an additional, adsorption-based desulfurization system to remove the sulfur from the JP-8 fuel.

Figure 5. Power dense SOFC stacks developed to support AFRL/Army TARDEC effort to demonstrate compact, light weight SOFC systems for mobile applications.

## NEXT GENERATION SOFC STACK DEVELOPMENT

As illustrated in the proceeding section, the weight of the SOFC stack is a major driver in the overall power density of an SOFC-based system. Though significant progress in designing and building ultra-compact SOFCs has been accomplished through the AFRL/TARDEC system development program, further progress is required to meet desired system-level power densities in excess of 200 W/kg, which require stack power densities >500 W/kg. In addition, compatibility with complex fuels, such as JP-8 and diesel, will require the ability to operate in a high sulfur environment. AFRL has sought to meet these challenges through a coordinated balance of contractual and in-house activities. The following discussion will describe some of the elements of AFRL's in-house strategies focused on meeting stack power densities of greater than 500 W/kg and sulfur tolerance in excess of 300 ppm in logistic fuel reformate, corresponding to 3000 ppm sulfur in the liquid fuel.

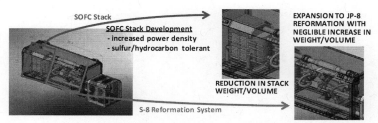

Figure 6. Primary technical challenges to achieving a JP-8 fueled, power dense SOFC system.

### Interfacial Modification

The first of these challenges is to increase the power density of the SOFC by almost 2X over the current AFRL/TARDEC program accomplishments. In order to achieve this level of stack performance, improvements in stack thermal management, gas manifolding, and current collection are all of critical concern, but the core of the issue is the cell level performance which is the area that AFRL has chosen to focus their in-house efforts. AFRL is actively exploring several in-house strategies focused on improving cell power density through development of novel cell processing methodologies. It is believed that development of these processes will allow greater control of the cell triple phase boundary regions permitting the ability to optimize active electrode functional layer design. During the past 3 years, AFRL has investigated the benefits of ink and aerosol jet deposition in order to effectively balance porosity, electrode surface area, and composition for additional design options at cell interfaces[14,15,16]. These additive manufacturing approaches differ from each other in the manner in which the depositing material is stabilized and deposited. In conventional ink jet printing, the depositing material remains in a stabilized suspension until a droplet reaches the substrate. Aerosol jet deposition, by contrast, uses either pneumatic or ultrasonic means to entrain the depositing material into a columnar gas stream which is deposited onto the substrate as a focused jet. Each of these approaches has relative advantages and can be used cooperatively as a means to produce thick ceramic films (>1 μm) and will be the subject of future publications[17].

Advanced deposition approaches, including ink and aerosol jet, offer several advantages when compared with traditional thick film processing strategies. Some of these advantages include the ability to deposit material masklessly, vary layer compositional properties and deposit highly repeatable layers. Maskless deposition is advantageous as it provides the ability to rapidly prototype or change layer properties based upon cell or stack requirements. As such, stacks from the same process line could be produced initially for higher power density operation, with larger current collecting layers and thinner electrolytes. This process could then be rapidly converted to produce cells optimized for longevity with more robust films or different material compositions. Figure 7 illustrates the ability of these methodologies to produce extremely thin or abrupt film contours without the use of a mask. For example, Figure 7a demonstrates the on-the-fly advantages of ink-jet deposition where two regions of YSZ electrolyte are deposited onto a nickel/YSZ cermet support, leaving a thin band of uncoated cermet remaining. Figure 7c illustrates a completed circular cathode layer which has been deposited on top of an thin electrolyte film without the use of a mask. In the case of aerosol jet, features as thin as 10 µm can be deposited on various surfaces allowing for deposition on even curved substrates. Figure 7d shows thin traces of silver which has been applied to the back of a LSM cathode. While these features are far larger than the 10 µm feature capability of this approach, the ability to masklessly write high aspect structures is illustrated.

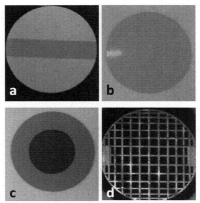

Figure 7. (a) Green NiO-YSZ anode support with an ink-jet applied electrolyte; (b) Fired NiO-YSZ anode support with crystalline YSZ electrolyte prepared by ink-jet; (c) Ink-jet prepared LSM cathode; (d) Aerosol-jet prepared silver current collector on LSM cathode.

Investigations are underway by AFRL to explore the use of these additive deposition techniques as a means for varying the compositional properties of layers. Through the use of these processing approaches, changes in material composition, porosity, and layer morphology can be achieved in the x, y, and z dimensions. The benefit to the cell is that it allows the designer extra degrees of freedom in the design of the critical functional (anode and cathode) layers thereby providing a tool to optimize a cell for a particular operating region. That is, cells designed for higher power density might have thinner functional layers with a higher density of catalytic sites as compared to cells designed for optimal efficiency. This is of particular interest to the DoD as the majority of fuel cell-based power systems are being designed for more commercial applications wherein extremely long life and low cost are preferential to performance. The ability to vary the cell design based on the needs of the customer makes these additive processing approaches desirable.

A secondary advantage to these approaches is that it has the potential to improve the structural integrity of the cell through improved balance between the thermal expansion properties of the different layers. The ability to vary cell composition in the axial dimension allows the design of compositionally graded interfaces where there is less of an abrupt interface between layers. In traditional cells, high regions of stress can occur between layers during thermal cycling due to differences in the different material coefficients of thermal expansion. This has been observed to cause catastrophic failures, especially along the interfacial regions between the layers which can manifest as delamination. In theory, a graded interface would allow this region of stress to be spread over a wider area decreasing the absolute level of strain seen at any single point in the interface region. While it is recognized that designing cells for improved structural integrity will likely be at odds to improving the performance, the use of these advanced deposition approaches could provide a design tool not currently available.

To test this capability, an experiment was conducted wherein both anode and cathode functional layers were compositionally graded. The anode functional layer was linearly varied from 100% YSZ directly adjoining the electrolyte to a 50% NiO/YSZ composition. The cathode functional layer (CFL) was also varied, but composition was changed from 100% YSZ at the electrolyte to 100% LSM over approximately 20 μm. Figure 8 shows the SEM image of this cell with the associated EDAXS mapping. As is observed, the desired compositional gradation appears to have been achieved with rapid variations produced for the cathode functional layers compared to a more gradual transition produced on the anode functional layer (AFL). It is also evident from the SEM that the electrode porosity appears to be sufficient even in proximity to the electrolyte and that the electrolyte layer appears to be continuous and defect free. The performance of this cell was not exceptional with a maximum power density of 0.25 W/cm$^2$ which is approximately 50% of the power density observed for cells produced by this method but with non-graded interfaces. However, the flexibility of the approach is still evident and yet to be optimized for performance.

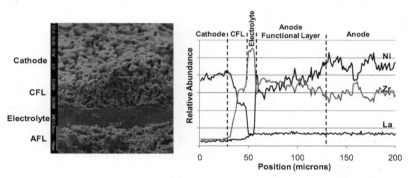

Figure 8. (a) SEM image of compositionally graded Ni-YSZ anode-supported SOFC cross-section; (b) EDAX data indicating compositional gradation of the cell.

While there are numerous challenges which must be overcome before this approach is suitable for high throughput manufacturing, several exciting fundamental questions remain. First, it must be established whether these deposition methods can be used to design cell interfaces to optimize for high performance or cell longevity. Secondly, the optimal design of each functional layer must be explored in terms of composition, porosity, and morphology. Lastly, the effective limits of cell dimensions must be explored to assess the scale-up potential of these techniques. In order to elucidate this point, very fundamental questions regarding the effective diffusional mean free path of ion fluxes within cell functional layers must be addressed and related to relevant cell parameters such as current density, reactant flow, etc. Furthermore, detailed surface science and redox kinetic measurements within the electrode functional layers must be performed in order understand the mechanism of the relevant redox reactions. It is these parameters which will be critical in the effective design of power dense cell interfaces.

Sulfur-Tolerant Anode Materials

A second challenge facing AFRL's efforts to develop a power dense, logistic fueled-capable SOFC is the development of cells and stacks which can permit operation with sulfur containing fuels. JP-8, the primary battlefield fuel, is a kerosene-type fuel which can contain as much as 3000 ppm sulfur in the liquid phase, which corresponds to ~300 ppm in the gas phase after fuel reformation. While most feedstocks are well below this limit, the USAF must be prepared to operate in any region and often depends on the host region to support fuel infrastructure[18]. As such, AFRL has established a basic research and development program focused on producing SOFCs which are tolerant to sulfur up to 300 ppm as $H_2S$ in the gas phase reformate.

Cell degradation due to sulfur poisoning is a complex phenomenon and addressing this degradation requires careful consideration of a multitude of factors including cell materials and cell design. In traditional cells, based on Ni-YSZ cermets, sulfur preferentially adsorbs on the catalytically active sites in the triple phase region of the anode functional layer and prevents adsorption and oxidation of fuel. While this deactivation is rapid it is also generally reversible once sulfur input to the cell is removed and oxide ion fluxes permit rapid oxidation of the sulfur as sulfur dioxide. However, for most metal-based anode formulations, the formation of a chemically bound sulfur species is spontaneous and facile resulting in the formation of a metal sulfide, if oxidation rates are insufficient to permit sulfur oxidation (e.g. low power density operation points). Metal sulfides, as nickel sulfide in standard anode formulations, are generally lower in melting point compared to the base metal. As such, the anode rapidly degrades due to a rearrangement of the interface, loss of available catalytically active surface and reduction in the structural properties of the anode.

In order to address this degradation, advancements in both material development and improvements in cell design are required. High surface area interfaces which permit the rapid oxidation of adsorbed sulfur species are being explored through the use of the ink delivery techniques described earlier. While maintaining a high current operating environment helps to improve cell sulfur tolerance, deactivation will ultimately occur without the use of advanced cell materials. One possible solution is to explore the use of doped, mixed-conducting metal oxides in lieu of traditional anode material sets to promote high sulfur tolerance. Perovskites, of the formula $ABO_3$, have shown promise due to their high stability in sulfur containing atmospheres. While these materials exhibit lower electronic and ionic conductivities than traditional anode material sets, doping of either A or B sites with elements of slightly dissimilar atomic radii can vary the conductivity of the resulting material[19]. As such, AFRL has sought to expand upon the work of others and to explore novel material formulations of both single and double-doped perovskites with the intent to produce a single component anode material with electrical conductivities sufficient for cell operation[13,20]. Figure 9 displays the Arrhenius-type behavior of several single and double-doped perovskite materials. As is observed from this plot, perovskites, either doped only at the A-site or through B-site doping exhibit increased conductivity compared to the undoped variants. For example, the doping of $SrTiO_3$ at the Sr-site with La increases the conductivity from 24 mS/cm at 750°C for the base perovskite to more than 380 mS/cm at La concentrations of 40%. This finding is consistent with earlier work performed for La doped titanates[21]. Lower concentrations of A-site dopant result in lower conductivity enhancements, but can provide benefit to the cell in terms of stability especially in cases where lanthanum diffusivity is problematic. Doping of the A-site also appears to have a desirable influence on the temperature dependence of conductivity. $SrTiO_3$ exhibits an activation energy of 19.5kJ/mol over the range of 400 to 800°C while lanthanum doped specimens exhibit 1.9 kJ/mol and 5.6 kJ/mol for the 20% and 40% lanthanum loadings, respectively. This observation is consistent with an increase in the mobility of oxide ions due to the introduction of the lattice defects due to the introduction of lanthanum. By contrast, doping of the B-site, can offer additional potential for improving the conductivity of perovskite materials. AFRL is actively studying material sets including $Sr_2MgMoO_6$ (SMMO), $Ba_2MgMoO_6$ (BMMO), and related materials as potential candidates for sulfur tolerant anodes. As their conductivities might suggest, these materials exhibit comparable or slightly improved performance compared with A-doped materials, but additional possibilities exist for improving the mixed conductivity performance of these materials through modification of cell structure.

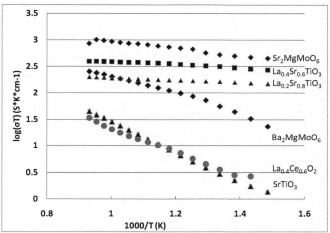

Figure 9. The conductivity of A and B-site doped perovskites. $La_{0.4}Ce_{0.6}O_2$ and $SrTiO_3$ are shown as a reference.

To evaluate the performance of these materials in a relevant environment, SMMO and BMMO were synthesized and deposited onto an electrolyte (LSGM) supported cell with a $La_{0.4}Ce_{0.6}O_2$ (LDC) buffer layer and Pt cathode. These cells were evaluated in hydrogen with an $H_2S$ concentration of 140ppm. Figure 10 shows the performance of the SMMO variant as a function of time. Upon addition of $H_2S$ into the cell fuel, an initial decrease in cell performance was observed but rapidly improved with sustained operation of more than 150 hours during which time only a moderate decline in cell voltage was measured. This minor drop in cell voltage was assumed to be associated with a normal aging of the cell typical of an unoptimized cell design. However, when $H_2S$ feed to the fuel stream was terminated a rapid increase in performance was observed suggesting that that no irreversible degradation of the material set had occurred.

To elucidate the mechanism of cell degradation, the cell cross-section was examined by SEM/EDAXS and the mapping of sulfur content in the anode, barrier, and electrolyte layers is presented as Figure 11. As observed, the SMMO and LSGM layers did not appear to show meaningful levels of sulfur content with maximum concentration of less than 2 wt% which is within the expected limits of error for EDAXS. The LDC barrier layer, by contrast, exhibited much higher levels of sulfur content suggesting sulfur has become irreversibly bound to the LDC surface or possibly incorporated into the LDC lattice. The Gibbs energy of formation of potential lanthanum and Ceria sulfide compounds would suggest that LDC is unlikely to decompose to form either LaS, CeS, $Ce_2S_3$, or $Ce_3S_4$ but a full examination of potential degradation mechanisms has not been done. Further analysis is underway to assess nature of these sulfur species present in the LDC layer and to clearly understand the potentially synergistic relationship between the SMMO anode/current collecting region with the LDC barrier layer.

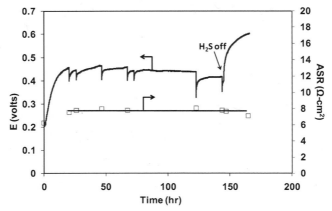

Figure 10. Galvanostatic test of a SMMO/LDC/LSGM/Pt cell measured at 750°C in $H_2$ with 140 ppm $H_2S$.

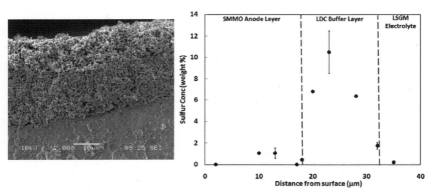

Figure 11. (a) SEM image of SMMO/LDC/LSGM/Pt cell cross-section; (b) Sulfur content in cell cross-section as measured by EDAXS.

CONCLUSION

AFRL is currently focused on the development of a power dense SOFC system capable of operating on military logistics fuels. To achieve this objective, AFRL/Army TARDEC are currently funding UTRC to develop a compact, lightweight SOFC system for UAV prime power and vehicle APU applications. The program has successfully demonstrated a 1.5 kW bench top prototype which operates at 50 W/kg on synthetic JP-8 fuel. Current efforts are focused on developing the next increment of this system with a power density goal of >100 W/kg and plans to perform a flight demonstration of this system as prime power for UAV in 2010. To achieve the power density goals and JP-8 operability required for an operational system requires further advances in SOFC stack technology including increased power density and tolerance to sulfur impurities. AFRL's in-house activities are exploring both these areas through direct-write SOFC fabrication approaches and sulfur tolerant anode material development. Utilizing direct-write approaches, AFRL has demonstrated the ability to functionally grade the composition between SOFC layers which may enable advanced cell designs focused on increased power density or increased efficiency and structural integrity. Furthermore, AFRL is exploring novel anode material formulations such as mixed-conducting perovskites. Through the addition of dopant materials, AFRL is working to increase the conductivity of the base perovskites in order the increase their suitability as an anode material. AFRL has demonstrated stable operation of these materials in $H_2$ with 140 ppm $H_2S$ for over 150 hrs of operation.

ACKNOWLEDGEMENTS

We would like to acknowledge our partners at the Army Tank and Automotive Research, Development and Engineering Center for their continuing support and cooperation. Also, we would like to acknowledge the team a United Technologies Research Center for their work on the SOFC power system development. In addition, we would like to recognize Dr. A. Mary Sukeshini for her leadership and work on direct-write SOFC fabrication and Thomas Howell for this work exploring anode materials for sulfur tolerant SOFCs.

REFERENCES

[1]Defense Energy Support Center Fact Book, www.desc.dla.mil (2008).
[2]H.H. Dobbs, T. Krause, R. Kumar, and M. Krumpelt, Diesel-Fueled Solid Oxide Fuel Cell Auxiliary Power Units for Heavy-Duty Vehicle, *4th European Solid Oxide Fuel Cell Forum*, Lucerne, Switzerland (2000).
[3]D.L. Daggett, S. Eelman, and G. Kristiansson, Fuel Cell APU for Commerical Aircraft, *AIAA Inter. Air and Space Symp. and Expo.*, Dayton, OH, **AIAA 2003-2660**, (2003).
[4]R.J. Braun, M. Gummalla, and J. Yamanis, System Architectures for Solid Oxide Fuel Cell-Based Auxiliary Power Units in Future Commercial Aircraft Applications, *J. Fuel Cell Sci and Tech.*, **6**, 031015-1 – 031015-10 (2009).
[5]Solid State Energy Conversion Alliance (SECA), U.S. Dept. of Energy, National Energy Technology Laboratory, www.netl.doe.gov/technologies/coalpower/fuelcells/seca/.
[6]W.A. Surdoval, Clean Economic Energy in a Carbon Challenged World, *International Pittsburgh Coal Conference*, Pittsburgh, PA (2008).
[7]E. Sun, J. Yamanis, L. Chen, D. Frame, J. Holowczak, N. Magdefrau, S. Tulyani, J. Hawkes, C. Haugstetter, T. Radcliff, and D. Tew, Solid Oxide Fuel Cell Development at United Technologies Research Center, *ECS Transactions*, **25(2)**, 77-84 (2009).
[8]J. Beals, R. Braun, F. Chen, P. Croteau, S. Emerson, J. Hawkes, C. Haugstetter, T. Radcliff, E. Sun, D. Tew, J. Yamanis, and N. Erikstrup, Development of a kW-Class Power Dense Solid Oxide Fuel Cell System, *Proc. of 43rd Power Sources Conf.*, 7-10 July, Philadelphia, PA, **Session 5.4**, 85 – 87 (2008).
[9]J. Hawkes, K. Centeck, M. Drejer-Jensen, S. Emerson, N. Erikstrup, T. Hale, C. Haugstetter, T. Junker, A. Kuczek, T. Lawson, R. Miller, T. Radcliff, M. Rottmayer, E. Sun, D. Tew, E. Wong, and J. Yamanis, A Power Dense 2 kW Logistics Fueled Solid Oxide Fuel Cell System for Mobile Applications, *Fuel Cell Seminar*, 16-19 Nov, Palm Spring, CA, **HRD41-3** (2009).

[10]N. Erikstrup, M. D. Jensen, M. R. Nielsen, T. N. Clausen, and P. Larsen, Power Dense SOFC Stacks for Mobile Applications, *ECS Trans*, **25(2)**, 207 - 212 (2009).

[11]M.J. Day, S.L. Swartz, L.B. Thrun, K. Chenault, and J.R. Archer, NexTech's Planar SOFC Technology, *Fuel Cell Seminar*, 16-19 Nov, Palm Spring, CA, **HRD24-4** (2009).

[12]M.R. Pillai, I. Kim, D.M. Bierschenk, and S.A. Barnett, Fuel-Flexible Operation of a Solid Oxide Fuel Cell with $Sr_{0.8}La_{0.2}TiO_3$ Support, *J. Power Sources*, **185**, 1086 – 1093 (2008).

[13] M. Gong, X. Liu, J. Trembly, and C. Johnson, Sulfur-Tolerant Anode Materials for Solid Oxide Fuel Cell Application, *J. Power Sources*, **168**, 289-298 (2007).

[14] D. Young, A.M. Sukeshini, R. Cummins, H. Xiao, M. Rottmayer, and T. Reitz, Ink-jet Printing of Electrolyte and Anode Functional Layer for Solid Oxide Fuel Cells, *J. Power Sources*, **184**, 191-196 (2008).

[15] A. M. Sukeshini, R. Cummins, T.L. Reitz and R.M. Miller, Ink-Jet Printing: A Versatile Method for Multilayer Solid Oxide Fuel Cells Fabrication, *J. Am Ceram. Soc.*, **92(12)**, 2913–2919 (2009).

[16]A. M. Sukeshini, R. Cummins., T. L. Reitz and R.M. Miller, Inkjet Printing of Anode Supported SOFC: Comparison of Slurry Pasted Cathode and Printed Cathode, *Electrochem. and Solid-State Lett.*, **12 (12)**, B176 – B179 (2009).

[17]A. M. Sukeshini, P. Gardner, T. Jenkins, R. Miller, and T. L. Reitz, Investigation of Aerosol Jet Deposition Parameters for Printing SOFC Layers, *8th International Fuel Cell Science, Engineering & Technology Conference*, Brooklyn, NY (2009), *Submitted*.

[18]D.D. Link, J.P. Baltrus and K.S. Rothenberger, Class- and Structure-Specific Separation, Analysis, and Indentification Techniques for the Characterization of the Sulfur Components of JP-8 Aviation Fuel, *Energy & Fuels*, **17(5)**, 1292-1302 (2003).

[19]H. Ullman and N. Trofimenko, Estimation of Effective Ionic Radii in Highly Defective Perovskite-type Oxides from Experimental Data, *Journal of Alloys and Compound*, **316**, 153-158 (2001).

[20]Y.H. Huang, R. Dass, Z.L. Xing, and J. Goodenough, Double Perovskites as Anode Materials for Solid-Oxide Fuel Cells, *Science*, **312**, 254-257 (2006).

[21]O.A. Marina, N.L. Canfield, and J.W. Stevenson, Thermal, Electrical, and Electrocatalytical Properties of Lanthanum-Doped Strontium Titanate, *Solid State Ionics*, **149**, 21-28 (2002).

# MICRO-TUBULAR SOLID OXIDE FUEL CELLS WITH EMBEDDED CURRENT COLLECTOR

Marco Cologna, Ricardo De La Torre, Vincenzo M. Sglavo
DIMTI, University of Trento
Trento, ITALY

Rishi Raj
University of Colorado
Boulder, CO, USA

ABSTRACT
In the present work a novel concept of micro-tubular solid oxide fuel cell is proposed. These cells have a sub–millimeter diameter. The metallic anode current collector, in the form of a wire forms a central, supporting core of this system. The micro–tubular SOFC is then fabricated by coating the wire with ceramic slurries that are sintered in succession. This paper reports the coating and sintering conditions that lead to a gas tight electrolyte and a porous inner anode architecture.

INTRODUCTION
Among the different solid oxide fuel cell (SOFC) architectures, micro-tubular is gaining increasing interest especially for portable applications. The major advantages in the *micro tubular* SOFC design, are (i) high volumetric power density compared to the standard tubular design (since the theoretical volumetric power density is inversely proportional to the electrolyte diameter), (ii) high thermal shock resistance, and (iii) rapid turn on/off capability[1-3] resulting from low thermal mass.

Tubular solid oxide fuel cells, having diameter greater than a few millimeters, are usually produced by extrusion or co-extrusion of the support electrode, which may be the anode or the cathode,[4-6] with the electrolyte. In some cases, an inner sacrificial core of a fugitive material (*e. g.* carbon black),[7] is used; the current collector is then inserted within the tube in a second step. Efforts have been made to reduce the total cell diameter to the sub-millimeter range. However, one of the limiting factors preventing the cells to be further scaled down are the technological challenges encountered in extruding such small tubes, and in inserting the inner current collector into such very small-diameter hollows.[2]

In the new process presented in this paper a thin ceramic electrolyte layers are deposited on to a metallic wire, which serves as the anode current collector, to construct a sub–millimeter tubular fuell cell system. A variety of well established technologies may be employed to deposit thin electrolyte layers in the tubular SOFC design. Examples are: conventional dip-coating from a slurry,[7-9] modified dip-coating,[10] sol-gel technique,[11, 12] electrochemical vapor deposition (EVD),[13] chemical vapor deposition from metal–organic precursors (MOCVD),[14, 15] physical vapor deposition (PVD),[16] electrophoretic deposition (EPD)[3] and a whole range of more or less conventional processing technologies.[17, 18] Among all the fabrication methods, it is usually recognized (although not unanimously) that colloidal processing is the most economical route for producing ceramics with good thermal and mechanical properties.[5, 19]

In this work we report the fabrication of a micro tubular solid oxide fuel cell with outside diameter in the sub-millimeter range. In this novel cell design, the support for the cell fabrication consists of a thin metallic wire, on to which the porous anode layer, the electrode and the cathode, are deposited in succession. Environmental impact and cost of the processing was minimized by using only water based technologies. The use of expensive equipment was avoided by using dip coating for cell fabrication. Anode and electrolyte were based on the state of the art materials. The anode was constructed from nickel oxide and 8mol% yttria stabilized zirconia (YSZ), while the electrolyte was made from YSZ. Slurries suitable for dip coating were developed in order to obtain a thin electrolyte coating onto the anode. The stability of the slurries was achieved by optimizing the powder / dispersant ratio via sedimentation tests. The organic binder amount was tailored in order to avoid drying defects. Systematic experiments, using dilatometry were used to optimize the best powders combination, as well as the suitable sintering temperature.

EXPERIMENTAL

The dispersant/powder ratio was optimized by sedimentation tests for each powder: yttria stabilized zirconia (TZ-8YS and TZ-8Y, Tosoh, Japan), nickel oxide ( J.T. Baker Inc., USA) and pore former (Graphite flake, median 7-10 $\mu$m, Alfa Aesar). Water dispersions with 4 wt% solid loading and variable dispersant concentration (ammonium polyacrylate, Darvan 821A, R.T. Vanderbilt Inc., USA) were mixed for two hours in a plastic jar containing YSZ grinding media and poured into graduated cylinders. The height of the sediment was measured periodically up to a maximum sedimentation time of 30 days. Suspension for dip-coating were prepared as follows: ceramic powders (YSZ in the electrolyte slurry and 58wt% NiO - 42wt%YSZ in the anode), dispersant and water were mixed in a high energy vibratory mill. Binders (either PEG600 and PEG20000, Fluka, or B1000 and B1014, Duramax B-1014 and d Duramax B-1000, Rohm and Haas) were added and the suspension was slowly mixed by a magnetic stirrer. The nickel support (Puratronic, Alfa Aesar) was coated with the anode slurry by dip-coating and drying and the electrolyte was applied in the same way on the outer surface of the dried anode. The cells were sintered at 1400°C after a slow de-binding step (1°C/min to 600).

Specific surface area (SSA) was determined by nitrogen adsorption (BET) method (ASAP 2010, Micromeritics, USA). Dilatometric analysis (Setsys Evolution, Setaram, France) were performed on cylindrical samples (around 5 mm in diameter and 4 mm in height) with heating rate of 5°C/min up to 1400°C and constant load of 5 g. The microstructure of the materials was analyzed by scanning electron microscope (SEM, JSM 5500, JEOL, Japan).

RESULT AND DISCUSSION

The optimal dispersant concentration, as estimated from the sedimentation tests, and the specific surface area (measured by nitrogen adsorption or as reported from the supplier datasheet) are reported in Table I. For the YSZ powders, the optimal dispersant concentration is proportional to the powder surface area, being 0.8 x $10^{-3}$ $ml_{dispersant}/m^2$, both for TZ-8Y or TZ-8YS. The same value is found for graphite, while for stabilizing a NiO suspension 1.4 x $10^{-3}$ $ml_{dispersant}/m^2$ are needed. Stable and well deflocculated colloidal suspensions, as required for dip-coating, were therefore employed.

Table I. Optimal dispersant concentration and specific surface area of the as received powders.

|  | NiO | TZ8-YS | TZ 8Y | Graphite |
|---|---|---|---|---|
| Optimal dispersant concentration [$10^{-3}$ $ml_{dispersant}/g_{powder}$] | 5 | 5 | 10 | 10 |
| Specific surface area [$m^2/g$] | 3.5[*] | 6[**] | 13[**] | 12[*] |

[*] Measured [**] From supplier datasheet.

Binder and plasticiser amount was tailored in order to prevent the coatings from cracking upon drying; this allowed to perform the drying step quickly at relatively high temperatures (80°C). Being the electrolyte layer thinner than the anode, and therefore the drying stresses lower, less organic phase was needed to avoid the drying cracks. Around 35 vol% organic binder was found to be sufficient in the electrolyte slurry, while more than 50 vol% had to be used in the anode slurry.

Dilatometric curves are reported in Fig. 1; the corresponding sintering rate is reported in Fig. 2. No or very little shrinkage due to de-binding was observed for all the electrolytes, while a shrinkage ranging from 2.2% to 3% was observed for all the anodes. This different behavior is explainable by noting that significantly lower binder amount was used in the electrolyte slurry. The anode shrinkage after debinding appears high, but it is worth noting here that the load, which is applied in the push rod dilatometer, although very small, can be responsible of magnifying the measured shrinkage in respect to a load-free dilatometer. Concerning the electrolytes, the high surface area one (TZ-8Y) is sintering at lower temperatures compared to the lower surface area (TZ-8YS) and to the mixture TZ-8YS/8Y (70/30 by weight), the temperature were the maximum sintering rate is observed being 1335°C, 1358°C and 1355°C, respectively. The same trend is observed for the anodes, the maximum sintering rate appearing at 1305°C for NiO - TZ-8Y, at 1340°C for NiO - TZ-8YS and at 1335°C for NiO - TZ-8YS/8Y (70/30). For both pure YSZ and NiO/YSZ mixtures, the use of a 70/30 ratio of low and high surface area powder is not significantly varying the sintering kinetics, compared to the low surface area powder only. The temperature of maximum sintering rate is always lower than 1400°C, this suggesting that a temperature of 1400°C can be sufficient for the cell sintering.

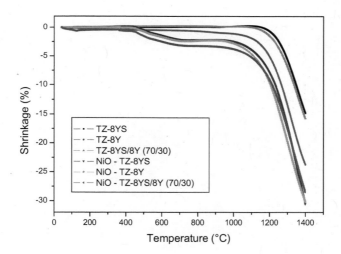

Fig. 1. Dilatometric curves of the anode and electrolyte samples. Heating rate 5°C/min.

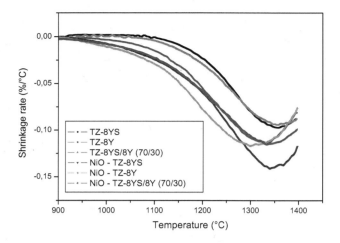

Fig. 2. Sintering rate of the anode and electrolyte samples. Heating rate 5°C/min.

Cells were built only with slurries containing NiO - TZ-8Y for the anode and TZ-8Y or TZ-8YS for the electrolyte. Various types of defects were observed in all samples after sintering at 1400°C half cells supported on straight Ni wire, no matter the starting powder, anode and electrolyte thickness and wire diameter. Such defects included longitudinal or transversal cracks and/or insufficient electrolyte densification. One example of such cracks is shown in Fig. 3. Such defects were attributed to the thermal and sintering stresses developed in the firing process.

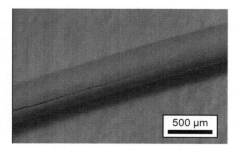

Fig. 3. Example of a defect found after sintering the anode and electrolyte coatings on a straight Ni wire.

In order to alleviate the overall stresses, the original cell design was modified and the Ni support wires were used after coiling them around a central core, therefore forming a spiral. The anode and electrolyte sintered on this less rigid support resulted in crack free layers. Half cells with crack free and dense electrolyte microstructure on porous anode could be easily obtained. The cracks observed in the coatings sintered on straight Ni wire could be related also to the superficial oxidation of the wire and the consequent volumetric expansion. Nevertheless, no cracks were observed in cells produced on the coiled wire support, this meaning that with such a geometry the coating can withstand the stresses associated to thermal expansion coefficient mismatch, constrained sintering and nickel volumetric expansion.

The best results in terms of electrolyte final density and absence of cracks were obtained with smaller wire and coil diameters. The electrolyte thickness was ranging from 10 μm to few tenths of micrometers, depending on the withdrawal speed and the anode support outer diameter. Half cell with a total diameter of 300 μm or even less were successfully obtained by coating nickel wires as thin as 50 μm coiled around a core of 100 μm in diameter. Depending on the coil spacing and anodic slurry viscosity, it was possible to coat the external part of the spiral only, allowing thus the direct creation of an internal hollow core for fuel flow, without the need of any sacrificial layer. The micrograph in Fig. 4 reports the cross section of a half-cell, where the central hollow, the support wire, the porous anode and the electrolyte are clearly visible.

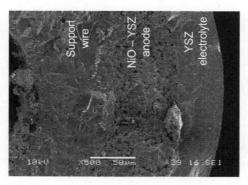

Fig. 4. Cross section of a tubular cell, as sintered.

Despite the volumetric power density is theoretically increased by decreasing the cell diameter, limits may exist in the miniaturization of a tubular cell. The first limit is the pressure drop along the length of the tube, which limits the cell length and the possibility to downscale the internal core diameter.[20] The second limitation is the cell / current collector ohmic resistance. As an example we can consider a cell with the following characteristics: 20 mm cell length, 50 μm Ni wire diameter, 200 μm coil pitch, 100 μm core diameter and anode layer with internal diameter 100 μm, external diameter 250 μm. As for the conductivity the pure Ni resistivity at 800°C of 15.2 μΩcm[21] and the porous Ni/YSZ anode conductivity ranging from 100 to 1000 S cm$^{-1}$can be considered.[22-24] The resulting resistance is equal to 4.3 Ω for the wire and 8.49 Ω or 84.90 Ω for the anode, depending upon the chosen conductivity. For a 100 μm diameter wire with a pitch between coils of 300 μm, the resistance is only 1.09 Ω. Not all the electrons, which are provided from the cell electrochemical reaction are running across the whole cell length, their average traveled distance being only one half of the cell length. Therefore, the real wire resistance is only one half of the above calculated one. Such a cell, under the hypothesis of a power density of 0.2 Wcm$^{-2}$ at 0.7 V would produce approximately 0.18 mA. The voltage drop caused by the anode and current collector ohmic resistance in the worst considered scenario (50 μm thick wire and 100 S cm$^{-1}$ anode conductivity) would be only 0.35 mV. One can conclude therefore that, although limits exist in scaling down the cell dimensions, the nickel wire, when properly engineered, can effectively reduce the total ohmic losses and decrease the limit on the minimum applicable diameter.

CONCLUSIONS

A new design for micro tubular SOFC has been presented. The cell is built around a thin metallic wire which serves both as a support for cell manufacturing, as well as a current collector. Inexpensive ceramic powder technologies were chosen for the cell fabrication. The use of a metallic spiral as a support, allowed the production of tubular cells with diameters as small as a few hundred micrometers, with a hollow core, a porous anode and a dense and crack-free electrolyte microstructure. The feasibility of a straightforward production route for sub-millimeter tubular SOFC with embedded current collector by means of very cheap technologies was thus demonstrated.

REFERENCES

[1] A. Dhir, K. Kendall, Microtubular SOFC anode optimisation for direct use on methane, *J. Power Sources*, 181, 297–303 (2008).

[2] S. C. Singhal, K. Kendall, High Temperature Solid Oxide Fuel Cells: Fundamentals, Design and Applications, Oxford, UK, Elsevier Ltd. (2003).

[3] P Sarkar, L. Yamarte, H. Rho, L. Johanson, Anode-Supported Tubular Micro-Solid Oxide Fuel Cell, *Int. J. Appl. Ceram. Technol.*, 4, 103–108 (2007).

[4] F. Tietz, H.-P. Buchkremer, D. Stöver, Components manufacturing for solid oxide fuel cells, *Solid State Ionics*, 152– 153, 373– 381 (2002).

[5] T. Alston, K. Kendall, M. Palin, M. Prica, P. Windibank, A 1000-cell SOFC reactor for domestic cogeneration, *J. Power Sources*, 71, 271-274 (1998).

[6] J–J. Sun, Y-H. Koh, W-Y. Choi, H-E. Kim, Fabrication and Characterization of Thin and Dense Electrolyte-Coated Anode Tube Using Thermoplastic Coextrusion, *J. Am. Ceram. Soc.*, 89, 1713–1716 (2006).

[7] T. Suzuki, T. Yamaguchi, Y. Fujishiro, M. Awano, Fabrication and characterization of micro tubular SOFCs for operation in the intermediate temperature, *J. Power Sources*, 160, 73–77 (2006).

[8] Z. Cai, T.N. Lan, S. Wang, M. Dokiya, Supported Zr(Sc)O$_2$ SOFCs for reduced temperature prepared by slurry coating and co-firing, *Solid State Ionics*, 152– 153, 583– 590 (2002).

[9] Y. Zhang, J. Gao, D. Peng, M. Guangyao, X. Liu, Dip-coating thin yttria-stabilized zirconia films for solid oxide fuel cell applications, *Ceramics International*, 30, 1049-1053 (2004).

[10] Y. Zhang, J. Gao, G. Meng, X. Liu, Production of dense yttria-stabilized zirconia thin films by dip-coating for IT-SOFC application, *J. Appl. Electrochem.*, 34, 637 – 641 (2004).

[11] X. Changrong, C. Huaqiang, W. Hong, Y. Pinghua, M. Guangyao, P. Dingkun, Sol-gel synthesis of yttria stabilized zirconia membranes through controlled hydrolysis of zirconium alkoxide, *J. Membr. Sci.*, 162, 181-188 (1999).

[12] K. Mehta, R. Xu, A. V. Virkar, Two-Layer Fuel Cell Electrolyte Structure by Sol-Gel Processing, *J. Sol-Gel Sci. and Tech.*, 11, 203–207 (1998).

[13] U. B. Pal, S. C. Singhal, Electrochemical Vapor Deposition of Yttria-Stabilized Zirconia Films, *J. Electrochem. Soc.*, 137, 2937-2941 (1990).

[14] K-W. Chour, J. Chen, R. Xu, Metal-organic vapor deposition of YSZ electrolyte layers for solid oxide fuel cell applications, *Thin Solid Films*, 304, 106-112 (1997).

[15] S. P. Krumdieck, O. Sbaizero, A. Bullert, R. Raj, YSZ layers by pulsed-MOCVD on solid oxide fuel cell electrodes, *Surf. Coat. Technol.*, 167, 226-233 (2003).

[16] T. H. Shin, J. H. Yu, S. Lee, I. S. Han, S. K. Woo, B. K. Jang, S-H. Hyun, Preparation of YSZ Electrolyte for SOFC by Electron Beam PVD, *Key Eng. Mat.*, 317-318, 913-916 (2006).

[17] J. Will, A. Mitterdorfer, C. Kleinlogel, D. Perednis, L.J. Gauckler, Fabrication of thin electrolytes for second-generation solid oxide fuel cells, *Solid State Ionics*, 131, 79–96 (2000).

[18]K. C. Wincewicz, J. S. Cooper, Taxonomies of SOFC material and manufacturing alternatives, *J. Power Sources*, 140, 280–296 (2005).

[19]F. F. Lange, "Powder Processing Science and Technology for Increased Reliability" *J. Amer. Ceram. Soc.*, 72 3-15 (1989).

[20]T. Yamaguchi, S. Shimizu, T. Suzuki, Y. Fujishiro, M. Awanoz, "Design and Fabrication of a Novel Electrode-Supported Honeycomb SOFC", *J. Am. Ceram. Soc.*, 92 S107–S111 (2009).

[21]D. R. Lide, CRC Handbook of Chemistry and Physics, 79th Edition, (1998-1999).

[22]A. Atkinson, S. Barnett, R. J. Gorte, J. T. S. Irvine, A. J. McEvoy, M. Mogensen, S. C. Singhal, J. Vohs, "Advanced anodes for high-temperature fuel cells" *Nature Materials*, 3 17-27 (2004).

[23]N. Sammes, Fuel Cell Technology Reaching Towards Commercialization. Germany, Springer, 2006.

[24]C. W. Tanner, A. V. Virkar, "A simple model for interconnect design of planar solid oxide fuel cells", *J. Power Sources*, 113, 44-56 (2003).

# DURABILITY IMPROVEMENT OF SEGMENTED-IN-SERIES CELL STACKS FOR SMALL SCALE SOFCs

T. Ito, Y. Matsuzaki

Product Development Dept., Tokyo Gas Co., Ltd.
3-13-1 Minami-Senju, Arakawa-ku, Tokyo 116-0003, Japan

## ABSTRACT
SOFC cells-stacks called "flat-tube segmented-in-series cells-stacks operable at intermediate temperatures" have been developed by a joint work of Tokyo Gas and Kyocera. The cells-stack is composed of multiple single-cells (electrolyte, anode, and cathode) and ceramic interconnects, placed on both sides of a flat-tubular substrate made of insulating material. Ceramic interconnects make the cells-stack more durable than other types of cellsstacks with alloy interconnects. The single-cells are electrically connected in-series on the substrate, so that a high-voltage is easily achievable. Durability tests under several conditions in electric furnaces were conducted in order to clarify the accelerating factors of deterioration. In the conditions, a particular test parameter was set to severe value. In the real system environment, the degree of variation of the parameters is larger than that in the standard condition in electric furnace. Therefore in order to estimate the degradation phenomenon in the real system, the durability tests were conducted also in the simulated system condition in the furnace with temperature distribution of 120 °C and reformed methane as a fuel. Then the cell-stacks were modified successfully to improve the durability with lower degradation rates under the several conditions.

## INTRODUCTION
Reduction of operation temperature and heat loss are important for thermally self-sustainable operation for small scale SOFC systems less than 1 kW aiming residential heat and power generation applications. The segmented-in-series type SOFC has advantages such as easy to get high voltage, simple connection between stacks with small current, etc. However this type of SOFC has been based on high temperature operations of around 1000 °C and has a complicated stack structure. Tokyo Gas and Kyocera are jointly engaged in the development of intermediate-temperature operable flat tubular segmented-in-series type cells-stack. The cells-stack structure is designed by Tokyo Gas, and the flat tube production technology is provided by Kyocera [1]. Tokyo Gas, Kyocera, Rinnai and Gastar have been co-operating in development of SOFC systems for residential applications using the segmented cells-stacks. Materials unique to the stack, such as electrical insulating materials for a support flat-tube, and ceramic material for interconnects, have been successfully developed. There are many requisitions for the support flat-tube material used here; high electrical resistivity for insulation of neighboring cells, appropriate porosity for rapid reactants diffusion, coefficient of thermal expansion (CTE) matching with other materials for thermal stress durability, high mechanical strength and acceptable cost. Catalytic activity for internal reformation is also important for moderation of demands to an external reformer. The cells-stack under development has many advantages such as reduced temperature operation by cell technology, high-voltage / low-current power generation by stack design, and lower in material cost of electrical insulating substrate compared to Ni based substrates for anode-supported cells. Another key advantage is that there is no need for alloy interconnects. This would make the cells-stack more durable than other types of cell-stacks having alloy interconnects. Long-term stability of the cells-stack is currently under investigation.

EXPERIMENTAL

The cells-stack design is schematically shown in Fig. 1. 16 single-cells are formed on one support flat-tube (8 single-cells on each side of the support flat-tube). Each single cell has Ni/8YSZ anode, Sr and Co doped-LaFeO$_3$ cathode, and 8YSZ electrolyte. Electrical current flows longitudinally at one side, and flows back at the other side. Fuel is supplied to each cell through ducts built inside the porous support flat-tube along the length of the flat-tube, and air is supplied to the cathode from outside of the flat-tube. An 8-YSZ film covering the flat-tube acts as electrolyte for each cell and separates the fuel from air.

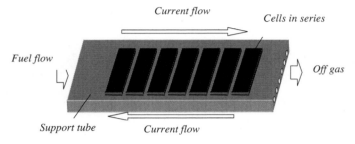

Figure 1. Schematics of overview of the segmented cells-stack.

Table 1 lists the materials for components of the cells-stack. Each component must have the proper stability (chemical, phase, morphological, and dimensional) in oxidizing and/or reducing environments, chemical compatibility with other components, and proper conductivity.

Table 1. Materials used in the flat-tubular segmented-in series type cells-stack.

| Component | Material |
|---|---|
| Anode | Ni-YSZ |
| Cathode | LSCF |
| Electrolyte | YSZ |
| Interconnect | Oxide |
| Substrate | Electrical insulator |

Cells-stacks with 16 single cells and/or short bundles which consist of 4 cells-stacks were manufactured and used for durability tests. Stack voltage was monitored as a function of operating time at a constant current density. Current was controlled by current pulse generator (Hokuto Denko Co. Ltd., HC-114). Ohmic resistance was obtained by separation from electrode polarization using a current interruption method.

Durability tests in an electric furnace were conducted under a standard condition as well as under a simulated system environment. In addition, the durability tests under controlled severe conditions, such as high current densities, large temperature distribution, and high fuel utilizations, were also conducted. Figure 2 shows the schematic image of the variation of test parameters such as temperature, fuel utilization, and so on, in the standard condition, the simulated system condition, the controlled severe condition as well as a real system condition.

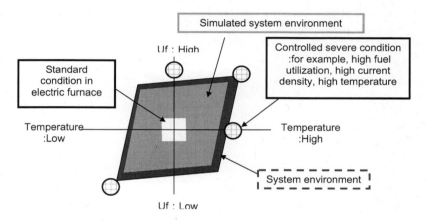

Figure 2. Schematic image of the variation of test parameters in the standard condition, the simulated system environment, the controlled severe condition, as well as a real system environment, where Uf indicates fuel utilization.

In the real system condition, the degree of variation of the parameters is larger than that in the standard condition in electric furnace. Therefore in order to estimate the degradation phenomenon in the real system, the durability tests were conducted in the simulated system condition in the furnace with temperature distribution of 120 °C and reformed methane as a fuel. Additionally in order to define accelerating factors of deterioration, the durability tests were conducted in the controlled severe conditions in which a particular test parameter was set to severe value. Figure 3 (a) and (b) show the external view of the short bundles installed in electric furnaces for the standard condition and the simulated system condition, respectively. In the case of the simulated system condition, the short bundle was set in a metallic chamber with thermal insulator inside.

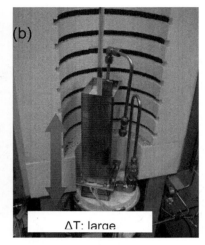

Figure 3. External view of the short bundles installed in electric furnaces for the standard condition (a), and the simulated system condition (b), respectively.

RESULTS AND DISCUSSION

In order to define accelerating factors of deterioration, the durability tests were conducted in the controlled severe conditions in which a particular test parameter was set to severe value. Two accelerating factors of deterioration of the cells-stack due to increase of the gas leakage in the cells-stack during operation have been found through the durability tests under the controlled severe conditions; one is high temperature and the other is high fuel utilization. Therefore cells-stack was modified to inhibit the increase of gas leakage during the operation [2].

Durability under the "standard" condition and simulated system environment

The cell-stacks with improved durability by modifying to inhibit the increase of gas leakage were tested. Figure 4 shows a short bundle's voltage as a function of operating time under the "standard" condition with a current density of 0.24A/cm$^2$, an average temperature of 775 °C having small temperature distribution less than 30 °C, and fuel utilization of 70%. The cells-stacks showed stable performance for experimental period up to 5000hrs. The degradation rates of the cells-stacks operated under the "standard" condition after 2000h and after 0h were found to be 0.31%/1000h and 0.70%/1000h, respectively. Additionally, the degradation rate of the cells-stacks operated from 0h to 4000h was found to be 0.87%/1000h. Figure 5 shows a short bundle's voltage as a function of operating time under the simulated system environment at a current density of 0.24A/cm$^2$ with fuel utilization of 70%, and relatively large temperature difference of around 120℃ between the top and bottom of the cell-stack, in which maximum and minimum temperatures were 820 °C at the top of the cell-stack and 700 °C at the bottom of the cell-stack, respectively. A thin line plotted in Fig. 5 indicates the voltage in the case of the "standard" condition shown above in Fig. 4. The degradation rate of the cells-stacks operated under the simulated system environment from 0h to 4000h was found to be 0.85%/1000h which was similar to the degradation rate of the cells-stacks operated under the "standard" condition from 0h to 4000h.

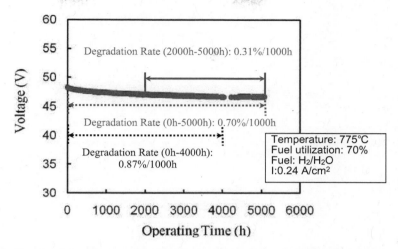

Figure 4. Short bundle's voltage as a function of operating time under the "standard" condition.

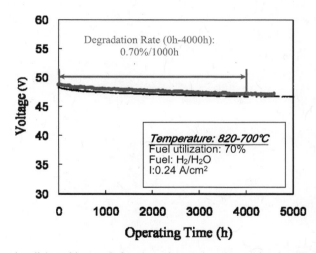

Figure 5. Short bundle's voltage as a function of operating time under the simulated system environment compared with voltage in the case of the "standard" condition plotted by a thin line.

Durability under the controlled severe conditions

In order to confirm the effects of the cells-stack modified to inhibit the increase of gas leakage during the operation, the durability tests were conducted in the controlled severe conditions in which a particular test parameter was set to severe value.

Figure 6 shows two short bundle's voltages as a function of operating time under the controlled severe conditions; one is under the high fuel utilization condition at a current density of $0.24A/cm^2$ with fuel utilization of 80% and a temperature of 775°C and the other is high temperature at a current density of $0.24A/cm^2$ with fuel utilization of 70% and relatively large temperature difference of around 120°C between the top and bottom of the cell-stack, in which maximum and minimum temperatures were 850 °C at the top of the cell-stack and 782 °C at the bottom of the cell-stack, respectively. A thin line plotted in Fig. 6 indicates the voltage in the case of the "standard" condition shown above in Fig. 4. The degradation rates of the cells-stacks operated under the high fuel utilization condition and the high temperature condition from 0h to 4000h were found to be 0.88%/1000h and 0.81%/1000h, respectively. These rates were similar to the degradation rate of the cells-stacks operated under the "standard" condition from 0h to 4000h. This result indicates that the high fuel utilization and the high temperature are no longer accelerating factors of deterioration for the cells-stack modified to inhibit the increase of gas leakage.

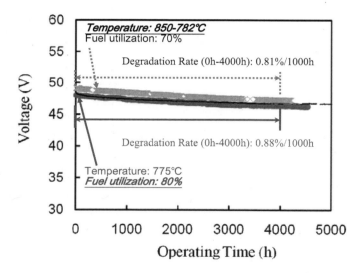

Figure 6.   Short bundle's voltage as a function of operating time under the controlled severe conditions compared with voltage in the case of the "standard" condition plotted by a thin line.

Durability of cells-stacks improved cathode

Durability of the cells-stacks modified to inhibit the increase of gas leakage was improved. Furthermore, the cells-stacks modified were improved about cathode. Figure 7 shows a voltage of short bundle improved cathode as a function of operating time under the "standard" condition with a current density of $0.24A/cm^2$, an average temperature of 775 °C having small temperature distribution less than 30 °C, and fuel utilization of 70%. A thin line plotted in Fig. 7 indicates the voltage of the cells-stacks unimproved cathode in the case of the "standard" condition shown above in Fig. 4. The degradation rate of the cells-stacks improved cathode under the "standard" condition from 0h to 4000h was found to be 0.21%/1000h which was considerably lower than that of the cells-stacks unimproved cathode under the same condition of Fig. 4. It is suggested that the durability of the cells-stacks was drastically increased by improving cathode.

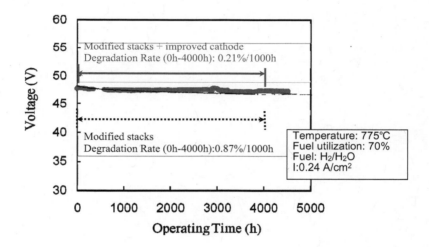

Figure 7. Short bundle's voltage of short bundle improved cathode as a function of operating time under the "standard" condition compared with voltage of the cells-stacks unimproved cathode in the same condition plotted by a thin line.

CONCLUSION

Segmented-in-series type cells-stack under development would have advantage in long-term stability because there is no need for alloy interconnects. The durability of the segmented type cell-stacks was investigated. Durability tests in an electric furnace were conducted under a standard condition as well as under a simulated system environment. In addition, the durability tests under controlled severe conditions were also conducted. Two accelerating factors of deterioration of the cells-stack due to increase of the gas leakage in the cells-stack unmodified during operation have been found through the durability tests under the controlled severe conditions; one is high temperature and the other is high fuel utilization. However these sever conditions are no longer acceleration factors of deterioration for the cells-stack modified to inhibit the increase of gas leakage during the operation. Cells-tacks modified under the "standard" condition at a current density of $0.24A/cm^2$ and temperature of 775 °C with small temperature distribution less than 30 °C showed stable performance for experimental period up to 5000hrs. The degradation rate of the cells-stacks modified from 2000h to 5000h and from 0h to 4000h was found to be 0.31%/1000h and 0.87%/1000h, respectively. Furthermore, the cells-stacks modified were improved about cathode. The degradation rate of the cells-stacks improved cathode under the "standard" condition from 0h to 4000h was found to be 0.21%/1000h which was considerably lower than that of the cells-stacks unimproved cathode under the same condition. It is suggested that the durability of the cells-stacks was drastically increased by improving cathode.

The cells-stacks were modified to inhibit the increase of gas leakage during the operation and improved cathode successfully, therefore the durability was drastically improved with lower degradation rates.

REFERENCES
[1]M. Koi, S. Yamashita, Y. Matsuzaki, 2007, Tenth International Symposium on Solid Oxide Fuel Cells (SOFC–X), ECS Transactions, 7(1), 235-243 (2007).
[2]Y. Matsuzaki, T. Hatae, S. Yamashita, 2009, Eleventh International Symposium on Solid Oxide Fuel Cells (SOFC–XI), ECS Transactions, 25(2), 159-166 (2009).

# PEROVSKITE MATERIALS FOR USE AS SULFUR TOLERANT ANODES IN SOFCS

X. Dong[a], P. Gardner[b], T.L. Reitz[b], F. Chen[a*]

[a] Department of Mechanical Engineering, University of South Carolina, Columbia, SC, USA
[b] Propulsion Directorate, Air Force Research Laboratory, Dayton, OH, USA
[*] Corresponding author. Email: chenfa@cec.sc.edu

## ABSTRACT

Sr- and Mn-doped $LaGaO_3$ ($La_{0.8}Sr_{0.2}Ga_{0.5}Mn_{0.5}O_{3-\delta}$, LSGMn) has been synthesized by solid state reaction (SSR) as well as glycine nitrate combustion (GNC) methods. XRD results show that pure single perovskite-phase can be obtained easily from the GNC method, while secondary phase exists from the SSR method even after repeated grinding and calcination of the precursors. The electrical conductivities of LSGMn material in air and in wet $H_2$ have been evaluated by four probe DC measurement for the sintered bar samples. Higher conductivities are observed for samples made from the GNC method compared with those from the SSR method. All perovskite-type fuel cells with $La_{0.8}Sr_{0.2}Ga_{0.83}Mg_{0.17}O_{3-\delta}$ (LSGM) pellet of the thickness between 300-400$\mu$m as electrolyte, LSGMn as anode and $La_{0.6}Sr_{0.4}Co_{0.2}Fe_{0.8}O_{3-\delta}$ (LSCF) as cathode have been fabricated and characterized. Preliminary fuel cell performance tests show that LSGMn is a good anode material for LSGM-based SOFCs. Although the electrolyte is relative thick, the highest power density at 800°C can reach around 400mW/cm$^2$ with $H_2$ as fuel and ambient air as oxidant. Further, LSGMn anode shows reasonable tolerance to sulfur poisoning when tested in $H_2$ with 100ppm $H_2S$.

## INTRODUCTION

Solid oxide fuel cells (SOFCs) have attracted a considerable attention due to their high energy conversion efficiency over conventional energy generation systems, low emission of green house gases and fuel flexibility compared with other types of fuel cells [1, 2]. Yttria-stabilized zirconia (YSZ) is the most commonly used electrolyte for SOFCs due to its desirable mechanical strength and chemical stability in both oxidizing and reducing atmospheres and its pure oxygen ion conductivity over a wide range of oxygen partial pressure. However, electrolyte-supported SOFCs using YSZ electrolyte typically operates around 1000°C since YSZ conductivity is relatively low at reduced temperatures [3]. The major research focus of recent times is to lower the operating temperature of SOFC in the intermediate temperature range (600-800°C) so as to make it commercially viable. This reduction in temperature is largely dependent on finding an electrolyte material with adequate oxygen ion conductivity at the intended operating temperature. Perovskite $LaGaO_3$ doped with Sr- and Mg-, $La_{0.9}Sr_{0.1}Ga_{0.8}Mg_{0.2}O_{3-\delta}$ (LSGM) has shown very high oxygen ion conductivity at intermediate temperature (600-800°C) over a wide range of oxygen partial pressure and is a promising electrolyte material especially for electrolyte-supported SOFCs [4-7].

Many current and future military applications require power generation technologies that can directly use logistic fuels and are fuel efficient and quiet. SOFCs offer superior efficiency potential and simpler system design. One of the critical challenges restricting broad implementation of SOFC-based power systems for military applications is the low stability of conventional Ni based cermet anodes to sulfur containing feedstocks [8-9]. Although Ni-based composite anode has good performance in $H_2$ as well as CO fuels and good current collection [10], it exhibits very limited tolerance to carbon deposition when using hydrocarbon fuels [11]. In addition, Ni-based anode has poor redox stability [12]. As nickel is such a good catalyst for hydrocarbon cracking, Ni cermet anode can only be utilized in hydrocarbon fuels if excess steam is present to ensure complete fuel reforming, diluting fuel and adding to system cost. Consequently, alternative sulfur tolerant anodes are needed to allow the direct use of practical fuels for military applications, without the need for extensive reforming and fuel conditioning.

Most oxides do not favor carbon-deposition and sulfur poisoning; therefore, replacing traditional Ni-based cermet anode with oxides would avoid carbon deposition and sulfur poisoning problems. Further, by judicious selection of the composition, oxides may possess redox stability. Good anode materials usually require both high electronic conductivity and ionic conductivity, and oxides with such properties are called mixed ionic/electronic conductors (MIECs). LaGaO3-based perovskite-type materials exhibit considerable mixed conductivity at elevated temperatures [6, 13]. As mentioned above, LSGM has been demonstrated to be a promising electrolyte material for intermediate temperature SOFCs. One of the basic requirements for being a good electrolyte is the redox stability. Consequently, LSGM shows excellent redox stability at typical SOFC operation conditions. It is reasonable to assume that doping certain transition metal ions in the B-site will create a LaGaO$_3$-based perovskite not only having mixed oxygen ionic and electronic conductivity but also maintaining redox stability. Further, LaGaO$_3$-based material is expected to possess sulfur tolerance since it has been shown to be promising anode material in SOFCs using H$_2$S-containing fuels [14]. In addition, Mn-doped LaCrO$_3$, (La$_{1-x}$Sr$_x$)$_{0.9}$Cr$_{0.5}$Mn$_{0.5}$O$_{3-\delta}$ (LSCM) has been reported to be a promising anode material with good chemical stability, good sulfur tolerance, reasonable catalytic activity for direct oxidation of hydrocarbon fuels and its catalytic properties seem more closely related to a manganite than a chromite [15]. Moreover, it has been reported that La$_{0.9}$Sr$_{0.1}$Ga$_{1-x}$Mn$_x$O$_{3-\delta}$ has good chemical stability as well as relatively high conductivity in H$_2$ atmosphere, making it an appropriate anode material for LSGM-based intermediate temperature SOFCs [16]. Therefore, La$_{0.8}$Sr$_{0.2}$Ga$_{0.5}$Mn$_{0.5}$O$_{3-\delta}$ (LSGMn) is chosen in this work as an alternative sulfur tolerant anode material for LSGM-based SOFCs. The use of LSGMn as anode materials for LSGM-based SOFCs offers at least two additional advantages. Firstly, it should be compatible both chemically and physically with LSGM electrolytes, minimizing interfacial reactions due to inter-diffusion or chemical reaction. Secondly, it should have thermal expansion coefficients similar to that of LSGM, minimizing thermally induced stresses at the interface during thermal cycling and, thus improving adhesion of the electrode to the electrolyte.

Processing condition is an important aspect that can greatly influence the electrical conductivity of the material. As the grain size, microstructure, phase purity and composition of the ceramic material are dependent on the processing conditions, optimization of processing conditions will certainly influence the electrical property of the material. LaGaO$_3$-based powders are typically synthesized by the solid state reaction (SSR) method which involves intimate mechanical mixing of oxides or carbonates and repeated grinding and heating cycles to achieve complete reaction between all reagents. Though this process is very simple, it requires multiple repetitions of prolonged thermal treatment and grinding. In addition, this technique produces large grains (1-10 μm) which are clearly a disadvantage as uncontrolled crystalline growth can occur, inducing chemical and grain-size non-uniformity [17]. The glycine nitrate combustion (GNC) method has been developed as a new synthesis route especially beneficial to the preparation of multiple component inorganic oxides [18-19]. This method offers several distinct advantages such as homogeneous mixing of several components at molecular or atomic levels, utilizing inexpensive reagents, and being able to produce ultra-fine powders with excellent phase purity.

The objective of this study is to explore synthesis of LSGMn through both SSR and GNC methods in an effort to demonstrate a sulfur tolerant anode material with equivalent electronic conductivity and catalytic activity of Ni-cermet anodes. The sulfur tolerance of LSGMn was evaluated by incorporating them into anode films on a LSGM electrolyte supported cell using La$_{0.6}$Sr$_{0.4}$Co$_{0.2}$Fe$_{0.8}$O$_{3-\delta}$ (LSCF) as the cathode material.

## EXPERIMENTAL

### Material Synthesis

LSGMn powders were prepared by both SSR and GNC methods. All the chemicals used are reagent grade. In the SSR method, the starting materials are $SrCO_3$, $La_2O_3$, $Ga_2O_3$ and $MnO_2$. $La_2O_3$ and $Ga_2O_3$ were heat-treated in a furnace at 950°C/1h to decompose trace carbonates before using. Stoichiometric $SrCO_3$, $La_2O_3$, $Ga_2O_3$ and $MnO_2$ powders were mixed through ball-milling overnight. The mixed powder was calcined at 1100°C for 5 h in air. After cooling down, the powder was grounded in an agate mortar and pestle and then ball-milled overnight. The powder was then calcined at 1200°C for 5h. Finally, the calcined powder was grounded and pressed into pellets and calcined again at 1400°C for 2h. The flow chart of this process is shown in Figure 1.

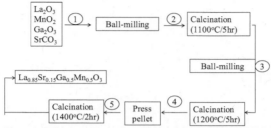

Figure 1. Solid-state reaction procedure for preparation of LSGMn

In the GNC method, the starting materials are La-, Sr, Ga- and Mn nitrates and glycine. The overall flow chart of this process is shown in Figure 2. Since La, Ga and Mn nitrates are moisture sensitive, it is very difficult to get an accurate measure of the nitrates using the powder form since the formula weight of the nitrates are not well defined due to the uncertainty of the number of water molecules absorbed in the nitrates. Therefore, the nitrates were dissolved in de-ionized water to make nitrate solutions, and the concentration of the solution was determined by a firing and weighing approach.

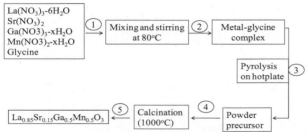

Figure 2. Glycine nitrate combustion procedure for preparation of LSGMn

To illustrate how the nitrate concentration was determined, lanthanum nitrate was taken as an example. 151g lanthanum nitrate hexahydrate powder was dissolved to make 500ml solution. 20ml lanthanum nitrate solution was taken into a dry clean 100ml ceramic crucible. The lanthanum nitrate solution was dried in an oven at 80°C to obtain lanthanum nitrate powder. The nitrate powder in the crucible was heated in a furnace at 1°C/min to 300°C, hold for 2 h, then 1°C/min to 800°C and hold for 2h. In this way, lanthanum nitrate was converted to $La_2O_3$, as verified by both thermogravimentric analysis and XRD tests. Three samples were used to derive an average value. The results are shown in Table 1. The concentration of lanthanum nitrate solution is calculated to be 0.7623 mol/L (in terms of the $La^{3+}$ concentration). In a similar way, the concentration of gallium nitrate solution was determined to be 0.3479 mol/L (in terms of the $Ga^{3+}$ concentration) and the concentration of manganese nitrate solution was 0.8523 mol/L (in terms of the $Mn^{2+}$ concentration).

Table 1.Firing and weighing results for lanthanum nitrate solution

|  | Empty crucible (g) | Crucible+ $La_2O_3$ (g) | $La_2O_3$ (g) |
|---|---|---|---|
| 1st time | 75.3071 | 77.7862 | 2.4791 |
| 2nd time | 64.6031 | 67.0914 | 2.4883 |
| 3rd time | 74.2924 | 76.7766 | 2.4842 |
| Average |  |  | 2.4839 |

In a typical process to produce about 15g LSGMn powders using the GNC method, 69.2ml lanthanum nitrate solution, 1.9707g strontium nitrate powder, 89.2ml gallium nitrate solution and 36.4ml manganese nitrate solution are taken to a 1000ml beaker. Glycine is added at a molar ratio of glycine to $NO^{3-}$ equal to 1:2, which is 12.47 in this case. De-ionized water was added to obtain a final volume of the nitrate glycine solution of about 400ml. The solution was heated in a hotplate while vigorously stirred. The temperature was kept at around 80°C. After stirring for 2hrs, the temperature of the solution was increased to about 90°C to accelerate the evaporation of the solution until a volume of about 80ml was left. The nitrate glycine mixture was now in a gel form, sticky and transparent. Further heating in the hotplate resulted in a self-ignition of the mixture, producing fluffy loose fine powders.

Thermogravimetric analysis (TGA) was performed on the gel as well as the combusted precursor powder from the GNC method on a TA Instruments TGA 2050 Thermogravimetric Analyzer. Data were collected between room temperature and 1000°C at a heating rate of 5°C/min under air atmosphere. Phase evolution of powders was determined by powder X-ray diffraction (XRD) analysis on a D/MAX-3C X-ray diffractometer with graphite-monochromatized CuKα radiation (λ=1.5418 Å), employing a scan rate of 3°/min in the 2θ range of 20 to 80°. The calcined LSGMn powders were pressed into bars and sintered at 1500°C for 5h.

Microstructure Characterization

The morphology or microstructure of LSGMn powders and solid bars were characterized by scanning electron microscopy (SEM, FEI Quanta 200) and transmission electron microscopy (TEM, Hitachi H-800, 200kV).

Electrical Conductivity

The conductivity of LSGMn bar was obtained from four-probe DC measurement (HP multimeter model 34401A). DC current was applied through the two end side Pt lead wires while the voltage was monitored by the two middle Pt lead wires. Platinum lead wires were attached on the bar samples by coating platinum paste and then firing at 1000°C for 1h. The LSGMn bar was placed in a tube furnace and the conductivity was first measured in air, then in $H_2$ (3% $H_2O$).

Fuel Cell Performance

To evaluate the performance of LSGMn anodes, single fuel cells were fabricated and measured. Both LSGM electrolyte material and LSCF cathode material were also synthesized by GNC method. The dense electrolyte pellet was obtained by sintering the LSGM pellets at 1500°C for 5h in air, and the thickness for the sintered pellets is about 300-400 μm. Both electrodes were applied on the two sides of the electrolyte pellet by screen-printing technique. The firing condition for the LSGMn anode was 1200°C/2h, while for LSCF cathode was 1100°C/2h. Platinum paste was applied on the surfaces of both electrodes as current collector, and platinum wires were attached to each electrode surface. The fuel cell performances in $H_2$ as well as in $H_2$ with 100ppm $H_2S$ were measured by 4-probe method with a Versa STAT 3-400 test system (Princeton Applied Research).

RESULTS AND DISCUSSION

Perovskite Phase Evolution

The XRD pattern of the LSGMn from the solid state reaction method after 1100°C/5h calcinations shows that there are impurity phases in the powder, as shown in Figure 3(a). The 1100°C/5h calcined powder was further ball-milled and calcined at 1200°C/5h. The power was then pressed into pellet and further calcined at 1400°C/2h. Even after such repeated calcinations, pure perovskite phase LSGMn could not be obtained, as shown in Figure 3(b). In comparison, the glycine nitrate combustion process is very effective in obtaining single perovskite LSGMn. The combusted powder shows single perovskite phase, as shown in Figure 3(c). After calcination at 1000°C/2h of the combusted powder, single perovskite phase LSGMn is unchanged, as shown in Figure 3(d).

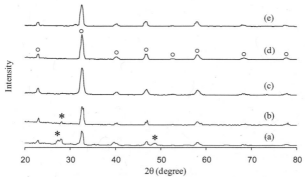

Figure 3. XRD patterns for LSGMn powders obtained at different conditions. (a) 1100°C/5h, SSR; (b) 1400°C/2h, SSR; (c) powder from combustion product, GNC; (d) 1000°C/2h of sample c; (e) 800°C/15h of sample d in 5%$H_2$-95%Ar. ° pervoskite phase; * impurity phase.

Compared with the traditional solid state reaction, by which the perovskite formation temperature is usually above 1100°C, and often with impure second phases, the glycine-nitrate combustion process can remarkably lower the phase formation temperature. In the glycine-nitrate process, the glycine functions not only as a fuel, but also as a complexing agent. One end of the glycine molecule, the amine group (-$NH_2$), can complex with the transitional metal ions, and the other end of the glycine molecule, the carboxyl group (-COO), can complex with alkaline earth metal ions. Because $La^{3+}$ ionic radius and chemical properties are similar to that of alkaline earth metal ions, it would also complex with carboxyl group. The complexation could prevent individual components from precipitation before combustion process, resulting in homogeneous products and lower phase formation temperature.

The thermogravimetric curve of the gel precursor from the glycine nitrate combustion method in flowing air is shown in Figure 4(a), which indicates that weight losses occur below 270°C, corresponding to the losses of water and decomposition of glycine and nitrates. The decomposition temperature of glycine is 233°C. The thermogravimetric curve of the powder from the combusted gel precursor in flowing air is shown in Figure 4(b), revealing of less than two percent weight losses. This weight loss is probably related to burning of carbon residual. Glycine serves as fuel for the combustion reaction, being oxidized by the nitrate ions. Stoichiometrically balanced [20], the exothermic reaction can be expressed as:

$$M(NO_3)_3 + M'(NO_3)_2 + NH_2CH_2COOH + O_2 \rightarrow M_2O_3 \text{ (M=La, Ga)} + M'O \text{ (M'=Mn, Sr)} + N_2 + NO_2 + CO_2 + H_2O \qquad (1)$$

Adjusting the ratio of glycine to nitrate, the combustion flame temperature can be adjusted. The glycine nitrate combustion process can produce an instant temperature as high as 1500°C [21]. Consequently, perovskite phase can be formed during the combustion process, as shown from the XRD result.

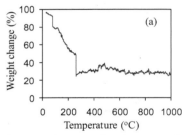 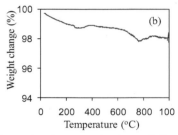

Figure 4. Thermogravimetric curves of (a) gel sample prior to combustion and (b) product after combustion.

Microstructure Evaluation

The morphology and particle size of the LSGMn powder from both SSR and GNC processes were characterized by electron microscope and the result is shown in Figure 5. SEM image (Figure 5b) shows that the LSGMn powder from GNC process is sponge-like and very loose. During the combustion process, large amount of gases ($N_2$, $NO_x$, $CO_2$ and $H_2O$) are evolved. This prevents agglomeration of the fine grains. At the same time, since the combustion process is very fast and lasts only about a few seconds, grain growth is greatly suppressed. Consequently, TEM analysis (Figure 5c) indicates that the individual LSGMn particle is nanocrystalline size (about 13nm in diameter). In comparison, the LSGMn powder from SSR method (Figure 5a) shows a particle size in the micron range, much larger than that of the GNC process. No conductive coating was applied to the LSGMn powders for the SEM analysis. It was observed that LSGMn powder from SSR showed much more charging effect than that of LSGMn powder from GNC, indicating that LSGMn powder from GNC method has higher electronic conductivity compared with that of LSGMn powder from SSR method.

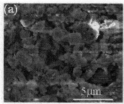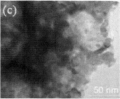

Figure 5. SEM images of LSGMn powder prepared from (a) SSR and (b) GNC. (c) TEM image of the LSGMn powder prepared from GNC.

Stability in Reducing Atmosphere

An important requirement for anode materials is good stability under reducing atmospheres. The chemical stability of the LSGMn was evaluated by heat-treating the 1000°C calcined LSGMn powder from GNC process (corresponding the sample in Figure 3d) in 5%$H_2$-Ar gas at 800°C overnight. XRD pattern (Figure 3e) of the LSGMn powder heat-treated in 5%$H_2$-Ar gas showed pure perovskite phase, indicating that the LSGMn possesses chemical stability in reducing environment at 800°C.

Electrical Conductivity

The conductivity of LSGMn from both SSR and GNC methods was evaluated using a four probe DC measurement. The samples are sintered LSGMn bars with a dimension of ~3.0cm in length, ~0.46cm in width (W) and ~0.18cm in thickness (T). The distance between the two middle Pt lead wires is 1.5cm (L). A typical sample and the lead wires attachment are shown in Figure 6. The sample was heated in an electric furnace and the conductivity was first measured in air. After the conductivity measurement in air, the sample was cooled down and then the conductivity was measured in wet $H_2$ ($H_2$ passing through water bubbler at room temperature, corresponding to ~3vol% $H_2O$). DC current was applied through the two end side lead wires while the voltage was monitored by the two middle lead wires.

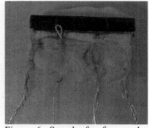

Figure 6. Sample for four probe DC conductivity measurement

Figure 7 shows Arrhenius plots of electrical conductivity of the LSGMn samples prepared from the two different methods measured either in air or in $H_2$. The samples made from SSR method have lower conductivity than that of the samples made from GNC, probably due to impurity phases in samples made from SSR. The conductivity increases with an increase in temperature and can be expressed in the Arrhenius equation [22]:

$$\ln(\sigma T) = \ln A - E_a/(RT) \qquad (2)$$

where A is the pre-exponential coefficient, $E_a$ is the activation energy, R is the universal gas constant and T is the absolute temperature. The activation energy for the electrical conduction is calculated from the slope of the curve. The activation energy in air is about 0.24eV for samples made from both GNC and SSR methods while in $H_2$ it is 0.16eV for samples from GNC and 0.35eV from samples from SSR. The activation energy for oxygen ionic conduction is typically around 1.0eV while the activation energy for electron hopping is about 0.1eV. The activation energy obtained in this study indicates that LSGMn shows a mixed ionic and electronic conductivity with predominantly electronic conduction.

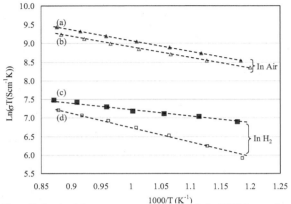

Figure 7. Conductivity vs temperature relationship for LSGMn samples. (a) and (c) are samples made from GNC method while (b) and (d) are samples made from SSR method.

Fuel Cell Performance

Single fuel cell was fabricated using LSGM as electrolyte, LSGMn as anode and LSCF as cathode. All the fuel cell materials were synthesized from GNC method. Shown in Figure 8 are the XRD patterns for LSGMn, LSGM and LSFC powders made from GNC method, indicating that all the three materials have pure perovskite phase. Therefore, the single fuel cell is an all pervovskite fuel cell.

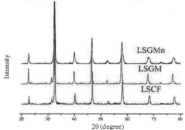

Figure 8. XRD patterns for single cell component powders obtained from GNC.

Figure 9 shows the fuel cell performance of the electrolyte-support single fuel cell at different operating temperatures. The maximum open cell potential is about 1.08 V at 650℃ in humidified $H_2$, and this value decreases with the increase in the cell testing temperature. The higher conductivity will result in lower resistance polarization and hence higher performance. The peak power density reached 400mW cm$^{-2}$ at 800°C, despite the relative thick electrolyte (400 μm). It is expected that further performance improvement can be obtained by decreasing the electrolyte thickness or optimizing the microstructure of the LSGMn anodes.

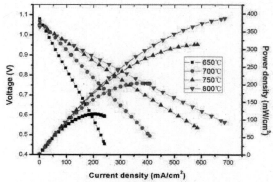

Figure 9. Fuel cell performance for the cell with LSGMn as anode

Shown in Figure 10 are the impedance spectra for the cell operated at different temperatures, indicating that the total impedance decreases with the increase of the cell operating temperature from 650°C to 800°C. The intercept of high-frequency arc with the real axis is related to the ohmic resistance of the cell, mainly the resistance of the electrolyte, while the difference between the real axis intercepts of the impedance is due to the interfacial polarization resistance. With the increase of the operating temperature, both the ohmic and interfacial polarization resistances decrease, resulting in improvement of the fuel cell performance, as shown in Figure 9.

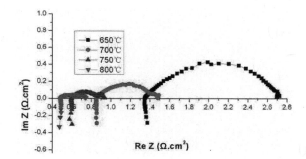

Figure 10. Impedance spectra of the cell operated at different temperatures

Figure 11 shows a typical current change as a function of time when the cell was operated at a constant voltage of 0.7V with either H$_2$ or H$_2$ with 100ppm H$_2$S as fuel gas in the anode while air as the oxidant in the cathode. Cell current decreased rapidly when 100ppm H$_2$S was introduced in the fuel gas stream, and the degradation rate slowed down with time and the cell performance was eventually stabilized. Upon removing $H_2S$ from the fuel gas stream, the current started to improve rapidly with time. However, the cell performance can not be fully recovered.

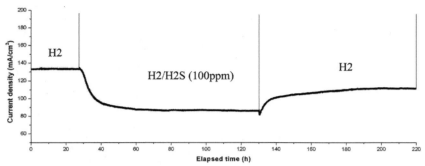

Figure 11. Long-term test of the cell with LSGMn anode in $H_2$ and $H_2$ with 100ppm $H_2S$

CONCLUSIONS
    LSGMn has been prepared from both solid state reaction and glycine nitrate combustion processes and the performance of the cell with the LSGMn as anode was investigated. Based on this study, the follow conclusions can be drawn:
    1. It is difficult to obtain single perovskite LSGMn phase from the solid state reaction approach. Even after calcinations at 1400°C, impurity phase still exists;
    2. Glycine nitrate combustion process is effective to synthesize single perovskite LSGMn with nanocrystal size;
    3. LSGMn has acceptable conductivity as electrode materials. At 800°C, the conductivity is >1S/cm in $H_2$ and >9S/cm in air.
    4. Although the electrolyte is relative thick, the highest power density at 800°C can reach around 400mW/cm$^2$ with $H_2$ as fuel and ambient air as oxidant.
    5. LSGMn anode shows reasonable tolerance to sulfur poisoning when tested in $H_2$ with 100ppm $H_2S$. However, the sulfur tolerance is still inadequate for military application where $H_2S$ concentration can be as high as 140ppm.

ACKNOWLEDGMENTS
X.D. and F.C. are grateful to the financial support of the Heterogeneous Functional Materials for Energy Systems, an EFRC funded by the U.S. Department of Energy, Office of Basic Energy Sciences under Award Number DE-SC0001061, the USC Future Fuels Initiative and NanoCenter, and the USC Promising Investigator Research Award. F.C. further acknowledges the support of the Summer Faculty Fellow Program from the Air Force Research Laboratory under Contract 09-S590-0004-03-C2.

REFERENCES
[1] A. Stambouli, and E. Traversa, Solid Oxide Fuel Cells (SOFCs): A Review of an Environmentally Clean and Efficient Source of Energy, Renew. Sustain. Energy Rev., 6, 433-55 (2002).
[2] O. Yamamoto, Solid Oxide Fuel Cells: Fundamental Aspects and Prospects, Electrochim. Acta, 45, 2423-35 (2000).

[3] F. Bozza, R. Polini, and E. Traversa, High Performance Anode-Supported Intermediate Temperature Solid Oxide Fuel Cells (IT-SOFCs) with $La_{0.8}Sr_{0.2}Ga_{0.8}Mg_{0.2}O_{3-\delta}$ Electrolyte Films Prepared by Electrophoretic Deposition, *Electrochem. Comm.*, **11**, 1680-83 (2009).

[4] R. Mitchell, Perovskites Modern and Ancient, Thunder Bay, Ontario: Almaz Press, 2002, ISBN 0-9689411-0-9.

[5] X. Lu, and J. Zhu, Effect of Sr and Mg Doping on the Property and Performance of the $La_{1-x}Sr_xGa_{1-y}Mg_yO_{3-\delta}$ Electrolyte, *J. Electroch. Soc*, **155**, B494-503 (2007).

[6] F. Chen, and M. Liu, Study of Transition Metal Oxide Doped $LaGaO_3$ as Electrode Materials for LSGM-Based Solid Oxide Fuel Cells, *J. Solid State Electrochem.*, **3**, 7-14 (1998).

[7] K. Huang, and J. Goodenough, A Solid Oxide Fuel Cell Based on Sr- and Mg-Doped $LaGaO_3$ Electrolyte: the Role of a Rare-Earth Oxide Buffer, *J. Alloy Comp.*, **303-304**, 454-64 (2000).

[8] Y. Matsuzaki, and I. Yasuda, The Poisoning Effect of Sulfur-Containing Impurity Gas on a SOFC Anode: Part I. Dependence on Temperature, Time, and Impurity Concentration, *Solid State Ionics*, **132**, 261-69 (2000).

[9] Y. Huang, R. Dass, and J. Goodenough, Double Perovskites as Anode Materials for Solid-Oxide Fuel Cells, *Science*, **312**, 254-7 (2006).

[10] W. Zhu, and S. Deevi, A Review on the Status of Anode Materials for Solid Oxide Fuel Cells, *Mater. Sci. Eng.*, **A362**, 228-39 (2003).

[11] J. Koh, B. Kang, H. Lim, and Y. Yoo, Thermodynamic Analysis of Carbon Deposition and Electrochemical Oxidation of Methane for SOFC Anodes, *Electrochem. Solid-State Lett.*, **4**, A12-5 (2001).

[12] M. Pihlatie, A. Kaiser, P. Larsen, and M. Mogensen, Dimensional Behavior of Ni-YSZ Composites during Redox Cycling, *J. Electrochem. Soc.*, **156**, B322-9 (2009).

[13] T. Ishihara, S. Ishikawa, K. Hosoi, H. Nishiguchi, and Y. Takita, Oxide Ionic and Electronic Conduction in Ni-doped $LaGaO_3$-based Oxide, *Solid State Ionics*, **175**, 319-22 (2004).

[14] S. Wang, M. Liu, and J. Winnick, Stabilities and Electrical Conductivities of Electrode Materials for Use in $H_2S$-containing Gases, *Journal of Solid State Electrochemistry*, **5**, 188-95 (2001).

[15] S. Tao, and J. Irvine, Discovery and Characterization of Novel Oxide Anodes for Solid Oxide Fuel Cells, *Chem. Record*, **4**, 83-95 (2004).

[16] Q. Fu, X. Xu, and G. Meng, Preparation and Electrochemical Characterization of Sr- and Mn-Doped $LaGaO_3$ as Anode Materials for LSGM-based SOFCs, *J. Mater. Sci.*, **38**, 2901-6 (2003).

[17] A. Sin, and P. Odier, Gelation by Acrylamide, a Quasi-Universal Medium for the Synthesis of Fine Oxide Powders for Electroceramic Applications, *Adv. Mater.*, **12**, 649-52 (2000).

[18] L. Chick, L. Pederson, G. Maupin, J. Bates, L. Thomas, and G. Exarhos, Glycine-Nitrate Combustion Synthesis of Oxide Ceramic Powders, *Mater. Lett.*, **10**, 6-12 (1990).

[19] L. Cong, T. He, Y. Ji, P. Guan, Y. Huang, and W. Su, Synthesis and Characterization of IT-Electrolyte with Perovskite Structure $La_{0.8}Sr_{0.2}Ga_{0.85}Mg_{0.15}O_{3-\delta}$ by Glycine-Nitrate Combustion Method, *J. Alloy Compds.*, **348**, 325-31 (2003).

[20] S. Jain, K. Adiga, and V. Verneker, A New Approach to Thermochemical Calculations of Condensed Fuel-Oxidizer Mixtures, *Combust. Flame*, **40**, 71-9 (1981).

[21] T. Peng, X. Liu, K. Dai, J. Xiao, and H. Song, Effect of Acidity on the Glycine-Nitrate Combustion Synthesis of Nanocrystalline Alumina Powder, *Mater. Res. Bul.*, **41**, 1638-45 (2006).

[22] M. Shi, N. Liu, Y. Xu, Y. Yuan, P. Majewski, and F. Aldinger, Synthesis and Characterization of Sr- and Mg-Doped LaGaO3 by using Glycine-Nitrate Combustion Method, *J. Alloy Compds.*, **425**, 348-52 (2006).

# PREPARATION AND CHARACTERIZATION OF LSCF
## $(La_{0.58}Sr_{0.4}Co_{0.2}Fe_{0.8}O_{3-\delta})$/GDC $(Ce_{0.8}Gd_{0.2}O_2)$ CATHODE FOR IT-SOLID OXIDE FUEL CELL

Na Li[1], Alevtina Smirnova[2,4], Atul Verma[2], Prabhakar Singh[2], Jeong-Ho Kim[3, a]

[1] Department of Mechanical Engineering, University of Connecticut
[2] Connecticut Clean Energy Engineering Center, University of Connecticut
[3] Department of Civil & Environmental Engineering, University of Connecticut
261 Glenbrook Rd., Storrs, CT 06269-2037, USA
[4] Department of Environmental Earth Science, Eastern Connecticut State University

## ABSTRACT

LSCF-GDC composite cathodes with different composition ratios were prepared and studied for potential application for SOFC with multifunctional graded cathodes. XRD showed no chemical reactions between LSCF and GDC at processing conditions considered. The mechanical properties such as the Young's modulus and hardness were measured by micro-and nano-indentation techniques. Thermal expansion behavior was recorded by a dilatometer. Electrical conductivity was measured by the four probe DC method. The effect of composition, sintering temperature and density on the mechanical and electrical properties is discussed. The measured experimental results are comparable to the literature and will be used as input for mechanical and electrical modeling, and as a reference for optimizing cathode materials.

## INTRODUCTION

Intermediate temperature (IT, 500-800°C) SOFC has gained a considerable attraction compared to a traditional high-temperature SOFC in that the reduced operation temperature allows low-cost metallic interconnects[1], helps to avoid material compatibility challenges at high temperature,[2] reduces sealing and thermal degradation problems,[3] and eventually accelerates the commercialization of SOFC technology. The overall cell performance, however, tends to decrease because of the reduced ionic conductivity of electrolyte and the increased polarization resistance of electrodes, especially on the cathode side.[1] Thus the development of higher-performance cathode is critical to overcome such technical barrier.

A commonly used means for improving the cathode performance is to add an ionically conducting second phase[45]. LSCF is one of the promising cathode materials for SOFC operated below 800°C. Impedance spectroscopy data for LSCF electrodes have shown lower interfacial resistance than conventional LSM electrodes[3]. The electrical conductivity of certain perovskite composition $La_{0.6}Sr_{0.4}Co_{0.2}Fe_{0.8}O_3$ can exceed 300S/cm[678]. This composition is also attractive because the high Fe content yields a low thermal expansion coefficient than low Fe compositions, better matching the low CTE of the electrolyte. Unfortunately, LSCF can not be used in conventional SOFCs because Co based cathode materials are chemically incompatible with YSZ electrolyte. They tend to form high resistance phase stronitium zirconate at the interface that deteriorates the cell performance.[910]The materials chemically compatible with Co-containing cathodes are doped ceria electrolytes, which posses a higher ionic conductivity than YSZ[11].

In the present work, the concept of functionally graded materials is used to enhance both electrochemical performance and mechanical durability of the LSCF-GDC SOFC. $La_{0.58}Sr_{0.4}Co_{0.2}Fe_{0.8}O_{3-\delta}$ and $Ce_{0.8}Gd_{0.2}O_2$ cathode composite with various volume fractions were used.

---

[a] Corresponding Author: Phone: (860) 428-2746, Fax: (860) 486-2298, Email: jhkim@engr.uconn.edu

Their electrochemical and mechanical properties were measured. The effect of composition, sintering temperature and density on the mechanical and electrical properties is discussed.

The ultimate goal of the work is to (1) characterize material properties of SOFC components (such as electrolyte, anode and cathode), especially composite cathodes with various volume fractions, to (2) perform three-dimensional thermal finite element analysis for a graded SOFC under thermal cycling, and to (3) perform electrochemical characterization of both multi-layered and conventional tri-layered SOFCs. The present material characterization study has provided basic framework for 3D finite element modeling that has recently demonstrated the enhanced performance of the graded SOFC compared to the tri-layered SOFC. This numerical observation will be validated with future experiments.

EXPERIMENTS

$La_{0.58}Sr_{0.4}Co_{0.2}Fe_{0.8}O_x$ with a specific area 12.2 $m^2/g$ was purchased from SEIMI Chemical Cooperation, and $Ce_{0.8}Gd_{0.2}O_2$ with a specific area 11.4 $m^2/g$ was purchased from DKKK in Japan. LSCF-GDC cathode powders were prepared by ball milling LSCF and GDC (Gd-doped Ceria) powders for 24h in ethanol to achieve good mixing. The weight percentages of LSCF in the mixing powders are 90%, 70% and 50%, respectively. Monolithic pellets and rectangular bars were obtained by dry pressing powders at about 200MPa and then sintered at 1200°C respectively in air for 2 hours with a heating rate 2°C/min. The linear shrinkage is calculated along the longitudinal dimension after sintering. The bulk density was determined by the Archimedes method with DI water as the immersing medium. Theoretical density was calculated using the lattice parameters obtained from diffraction analysis. Phase characterization was determined by XRD analysis using Cu Kα radiation (Bruker D5005 Advance X-Ray Diffractometers).

The sintered bars were about 3x4x50mm in dimension; Pellets were about 10mm in diameter and 2mm in thickness. Rectangular samples were used for thermal expansion tests and electrical conductivity tests. Thermal expansion properties were measured using the NETZSCH 402PC dilatometer in air over a range from the room temperature to 1000°C with a heating rate 3°C/min. Standard alumina rod was used for calibration. The average thermal expansion coefficient (α) was calculated from the expansion curve using

$$\alpha = \frac{1}{L_0} \frac{\Delta L}{\Delta T} \tag{1}$$

where $L_0$ is the initial length of the sample, $\Delta L$ is the sample length change and $\Delta T$ is the range of temperature variation. Electrical conductivity were measured using the standard DC four probe technique with a Keithley 2440 sourcemeter in a tube furnace in air upon cooling from 800°C to 200°C. Two inner platinum wires acted as current contacts and the other outer two wires acted as the voltage contacts. They were attached to the rectangular bars by painting platinum paste along the circumstance. A constant current was applied to the current wires and the voltage response on the voltage wires was recorded. The conductivity was determined by from a set of I-V values by taking

$$\sigma = \frac{L}{A} \times \frac{dI}{dV} \tag{2}$$

The measurements were performed at low V and I values to ensure the validity of Ohm's law.

The pellets were mounted using Buehler Epo-Thin low viscosity epoxy resin and hardener, and cured overnight at room temperature. Then the mounted samples were grounded by sand paper step by step from 180, 320, 400, 600 and 1200grit, finally polished down to 1um by diamond slurry.

Vickers micro-hardness test was done by employing Leco DM-400 FT hardness tester using 300g load held for 30s. At least 10 micro indents were made for each sample. The diagonal lengths of the indents were measured using an objective lens of 60X and the corresponding Vickers hardness was directly reported. Indentations were made in different areas and at least 50um distance between two indents to make the data more representative. Nano Indentation test was done using MTS Nano Indenter XP. The Berkovich hardness (H) and the elastic modulus (E) were calculated by software Testworks 4. All pellets were tested at room temperature with 50g applied load held for 30s. All the data were recalculated using Frame Stiffness 3.9E+6. The indents after micro- and nano- hardness tests were recorded using optical microscopy. The elastic modulus was calculated according to equation

$$\frac{1}{E_r} = \frac{1 - v_{indentor}^2}{E_{indentor}} + \frac{1 - v_{sample}^2}{E_{sample}} \tag{3}$$

where, $v_{indenter}$ and $v_{sample}$ are the Poisson ratios and $E_{indenter}$ and $E_{sample}$ are the elastic modulus of the diamond indenter and the sample, respectively[12]. The elastic modulus of the sample was calculated using $v_{indenter}$ =0.07 and $E_{indenter}$ =1140 Pa for the diamond indenter and assuming $v_{sample}$ =0.3[13].

RESULTS AND DISSCUSSION

The XRD patterns of the selected LSCF-GDC composite obtained after being sintered at 1000°C, 1100°C and 1200°C for 2hs are shown in Fig. 1. LSCF representing the Perovskite phase structure and GDC representing the fluorite structure are detected in the composites, and no second phase appeared during sintering process. It may be concluded that there is no chemical reactions between LSCF and GDC; they are chemical compatible up to 1200°C. The LSCF-GDC composite can be chosen as cathode materials and would not react with GDC electrolyte or interlayers.

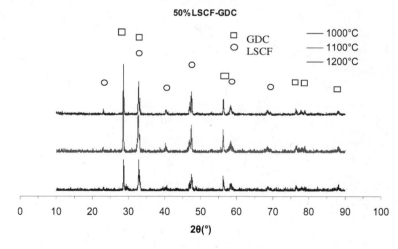

Fig.1 XRD patterns of 50wt% LSCF-GDC composite after sintering at different temperatures

Theoretical densities were 6.35g/cm$^3$ for LSCF and 7.31g/cm$^3$ for GDC, based on XRD lattice parameters. Theoretical densities of LSCF-GDC composites were calculated according to Vegard's law[14]. Fig. 2 shows the densification behavior of LSCF-GDC powders pressed at same pressure and sintered at same temperature 1200°C. All samples have high densities which are greater than 94% of theoretical densities. With the addition of LSCF content, the density tends to decrease, but not in a linear fashion.

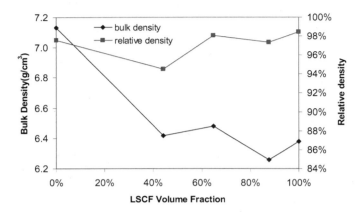

Fig.2 Densification characteristics of GDC, LSCF, and LSCF-GDC composites sintered at 1200°C

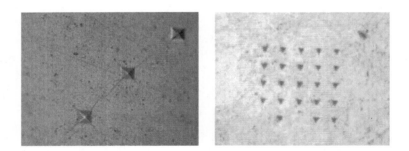

Fig.3 Optical microscopy of micro (left) and nano (right) indents

Fig. 3 shows typical micro and nano indent images. The four-sided micro indents are around 22μm in diagonal length and in spacing 50μm from each other. The triangular nano indents were distributed in a 5x5 pattern with 15μm distance between every two indents. The pyramid shape of the Berkovich indenter provides a sharper point than the four-sided Vickers geometry, therefore ensuring a more precise control over the indentation process. Fig. 4 shows the load-displacement curves of nano indentation tests. There is no pop-in formation that would indicate larger inhomogeneities in the material.[15] Upon the same loading, the larger penetration depth means the material matrix is weaker and deforms more. The initial slope of unloading segment was used to calculate the elastic modulus. The steeper the unloading curve, the bigger of elastic modulus. Moreover, the mechanical properties do not change significantly with the considered compositions.

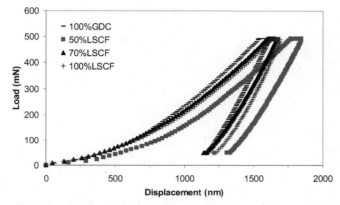

Fig.4 Nano-indentation load-displacement curves for different samples

The calculated elastic modulus and hardness data are presented in Fig.5 and Fig.6, and showed similar observation. Furthermore, the relative density has a trend similar to the modulus change with LSCF contents. Young's modulus is insensitive to flaw size and is a measure of bonding at an atomic level, but porosity lowers the Yong's modulus of ceramic materials[14]. The modulus and density results of boron suboxide are successfully fitted using linear or power law[16]. Both the micro and nano hardness data strongly relate to the change in density. For example, Kim[17] and Dianying[18] studied the relationship of sintering density and hardness of ceramic materials, and reported that hardness increased with the increased density. It is observed in the present study that the elastic modulus and hardness significantly depend on the density. Mechanical properties are shown to have a large deviation from sample to sample on the same composition.

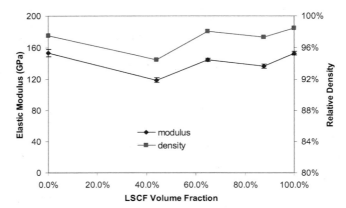

Fig.5 Elastic modulus (by nano indentation) and the relative density (by Archimedes's method) for samples with different LSCF volume fractions

As shown in Fig. 5, samples of 100% LSCF and 100% GDC in this experiment showed similar modulus and hardness values. The variation of the modulus is due to the density rather than LSCF volume fraction. In addition, Fig 6 shows that micro-hardness is lower than nano-hardness, since micro-indenter has a larger contact area than nano indenter and has more chance to expose to defects and pores.

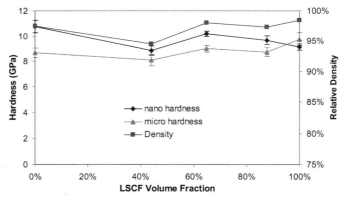

Fig.6 Micro- and nano- Hardness and the relative density for samples with different LSCF contents

Table 1 Density, electrical conductivity and thermal expansion data for LSCF-GDC samples

| LSCF weight content | LSCF molar content | Linear shrinkage | Apparent porosity | Bulk Density (g/cm³) | Theoretical Density (g/cm3) | Relative Density | TEC (K⁻¹/10⁻⁶) 20°C- | | Conductivity (S/cm) | |
|---|---|---|---|---|---|---|---|---|---|---|
| | | | | | | | 700°C | 1000°C | 600°C | 800°C |
| 0% | 0.0% | 14.60% | 3.40% | 7.13 | 7.31 | 97.5% | 12.0 | 12.7 | - | - |
| 50% | 44.1% | 17.10% | 1.90% | 6.42 | 6.79 | 94.5% | 13.9 | 15.7 | 140 | 112 |
| 70% | 64.8% | 16.90% | 4.10% | 6.48 | 6.61 | 98.0% | 14.1 | 16.1 | 251 | 205 |
| 90% | 87.7% | 17.70% | 2.80% | 6.26 | 6.43 | 97.3% | 13.9 | 16.2 | 324 | 271 |
| 100% | 100.0% | 17.20% | 3.70% | 6.25 | 6.35 | 98.5% | 14.6 | 16.9 | 366 | 314 |

All the samples sintered at 1200°C for 2 hours were measured for density, conductivity, and thermal expansion coefficients, and the experimental data as summarized in Table 1. They are comparable to the literature[7192021222324]. The linear thermal expansion of LSCF-GDC composite cathode with 0, 50, 70, 90, and 100wt% LSCF are shown in Fig.7 as a function of temperature. As expected, the percentage thermal expansion increases with temperature for all the samples. The slope for the LSCF-GDC samples also increases with increasing LSCF content. Among these, three samples the expansion is found to be lowest for the GDC electrolyte. The doped samples had closer thermal expansion properties to one another than pure LSCF and GDC samples. The TECs from room temperature to 800°C can be read from Fig.8: GDC 12.3 x10⁻⁶/°C, 50% LSCF 14.6 x10⁻⁶/°C, 70% LSCF 14.8 x10⁻⁶/°C, 90% LSCF 14.6 x10⁻⁶/°C, LSCF 15.5 x10⁻⁶/°C, respectively. It can be concluded that the thermal expansion coefficients increase with the increase of the LSCF content. Matching TECs in the SOFC components is critical, as small difference in the TEC of the cell components can produce larger thermal stresses during single cell fabrication and fuel cell operation. Multilayered cathode with graded composition of LSCF and GDC reduces the mismatch between GDC electrolyte and LSCF cathode, and increase the durability of fuel cells in operation.

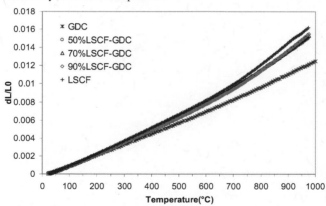

Fig.7 Comparison of linear thermal expansion of LSCF/GDC composite with various LSCF contents

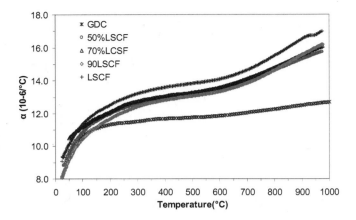

Fig.8 Thermal expansion coefficients of LSCF/GDC composite, and pure LSCF and GDC in air

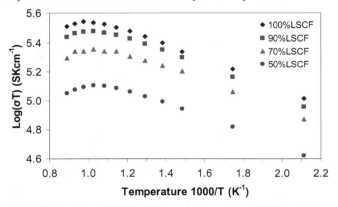

Fig.9 Conductivity of LSCF-GDC composites and pure LSCF cathodes

Fig.9 shows Log ($\sigma$T) of LSCF and LSCF-GDC bars as a function of reciprocal temperature. Clearly, the Log ($\sigma$T) curve has good linear dependence of reciprocal temperature, which is consistent with the small polaron conduction mechanism.

$$\sigma = \frac{c}{T}\exp(-\frac{E_a}{K_B T}) \tag{4}$$

where c is a constant, $E_a$ is the activation energy, k is the Boltzmann constant, and T is the absolute temperature. The conductivity of pure LSCF was measured as high as 366S/cm. With the addition of GDC content, the conductivity decreased accordingly. The electrical conductivity of all compounds increased with temperature up to about 600°C. At higher temperatures a steep conductivity decrease was observed. The conductivity decrease above 600°C is attributed to the loss of lattice oxygen[16 25].

CONCLUSIONS

LSCF and GDC composites with different compositions were prepared and the relative density of 94%-98% was achieved after sintering at 1200°C for 2hs. XRD analysis showed no chemical reaction between two phases. Thermal expansion behavior, electrical conductivity and mechanical properties were examined for these materials. Thermal expansion coefficient was in a range 12.3~15.5x10$^{-6}$/°C from room temperature to 800°C. Electrical conductivity was from 140 to 366 S/cm at 600°C. TECs and electrical conductivity increased with addition of LSCF contents in composite. Elastic modulus was 119~153 GPa, nano hardness 8.9~10.8 GPa and micro hardness 8.1~9.8 GPa. Hardness and elastic modulus depended more on density than LSCF composition. All the measured data are comparable to literature and fracture toughness will be measured for various LFCG-GDC compositions. The measured properties will be used for three-dimensional thermal finite element analysis for a graded SOFC under thermal cycling. This 3D modeling will be validated with future experiments.

ACKNOWLEDGEMENTS

JHK acknowledges the financial support from the National Science Foundation (NSF) under the Faculty Early Career Development (CAREER) Grant CMMI-0546225 (Materials and Surface Engineering Program). We'd also like to acknowledge the technical support from the C2E2 (Center for Clean Energy Engineering) and the IMS (Institute of Materials Science) for our experimental program. Also appreciated is the invaluable technical assistance of Prof. Rainer Hebert and his graduate student Ms. Girija Marathe in micro-and nano-indentation experiments. Any opinions expressed herein are those of the writers and do not necessarily reflect the views of the sponsors.

REFERENCES

[1] S. P. Jiang and W. Wang, Fabrication and performance of GDC-impregnated (La,Sr)MnO3cathodes for intermediate temperature solid oxide fuel cells, Journal of the Electrochemical Society 152(7) A1398-A 1408 (2005)

[2] Z. Duan, M. Yang , A. Yan, Z. Hou, Y. Dong, Y. Chong, M. Cheng, W. Yang, Ba0.5Sr0.5Co0.8Fe0.2O3−δ as a cathode for IT-SOFCs with a GDC interlayer, Journal of Power Sources 160 (2006) 57-64

[3] E. Perry Murray, M. J. Sever, S.A. Barnett, Electrochemical Performance of LSCF-GDC composite cathodes, Solid State Ionics 148 (2002) 27-34.

[4] Y. Matsuzaki and I. Yasuda, Electrochemical properties of reduced-temperature SOFCs with mixed ionic–electronic conductors in electrodes and/or interlayers, Solid State Ionics 152-153(2002) 463-468

[5] Y. Leng, S. H. Chan, Q. Liu, Development of LSCF–GDC composite cathodes for low-temperature solid oxide fuel cells with thin film GDC electrolyte, Internati Onal Journal of Hydrogen Energy 33 ( 2008 ) 3808 – 3817.

[6] L.-W. Tai, M.M. Nasrallah, H.U. Anderson, in: S.C. Singhal, H. Iwahara (Eds.), Proc. 3rd Int. Symp. Solid Oxide Fuel Cells, The Electrochemical Society Proceedings Series (1999) p. 241, Pennington, NJ.

[7] Anthony Petric, Peng Huang and Frank Tietz, Evaluation of La–Sr–Co–Fe–O perovskites for solid oxide fuel cells and gas separation membranes, Solid State Ionics 135 (2000), 719-725

[8] G. CH. Kostogloudis, Ch. Ftikos, properties of A-site-deficient La$_{0.6}$Sr$_{0.4}$Co$_{0.2}$Fe$_{0.8}$O$_{3-\delta}$ -based perovskite oxides, Solid State Ionics 126 (1999) 143-151

[9] A. Esquirol; N. P. Brandon; J. A. Kilner; M. Mogensen, Electrochemical characterization of $La_{0.6}Sr_{0.4}Co_{0.2}Fe_{0.8}O_3$ cathodes for intermediate-temperature sofcs, Journal of the Electrochemical Society 151(11) A1847-A1855(2004)

[10] H.Y. Tu, Y. Takeda, N. Imanishi and O. Yamamoto Solid State Ionics 117 (1999), p. 227.

[11] K. R. Reddy, K. Karan, Sinterability, Mechanical, Microstructural, and Electrical Properties of Gadolinium-Doped Ceria Electrolyte for Low-Temperature Solid Oxide Fuel Cells, Journal of Electroceramics, 15, 45–56, 2005

[12] Y. Wang, K. Duncan, E. D. Wachsman, F. Ebrahimi, The effect of oxygen vacancy concentration on the elastic modulus of fluorite-structured oxides, Solid State Ionics 178 (2007) 53–58

[13] Y.-S. Chou, J. W. Stevenson, T. R. Armstrong, and L. R. Pedernson, Mechanical properties of $La_{1-x}Sr_xCo_{0.2}Fe_{0.8}O_3$ mixed-conducting perovskites made by the combustion synthesis technique, J. Am. Ceram. Soc. 83960 1457-1464 (2000)

[14] R. A. Cutlera and D. L. Meixner, Ceria–lanthanum strontium manganite composites for use in oxygen generation systems, Solid State Ionics 159 (2003) 9 – 19

[15] A. C. Fischer-Cripps, Nanoinentation, second edition, Springer, 2004

[16] D. Music, U. Kreissig, ZS. Czig´any, U. Helmersson, J.M. Schneider, Elastic modulus-density relationship for amorphous boron suboxide thin films, Appl. Phys. A 76, 269–271 (2003)

[17] S. W. Kim and K. A. R. Khalil, High-Frequency Induction Heat Sintering of Mechanically Alloyed Alumina–Yttria-Stabilized Zirconia Nano-Bioceramics, J. Am. Ceram. Soc., 89 [4] 1280–5 (2006)

[18] D. Chen , E. H. Jordan, M. Gell, and X. Ma, Dense Alumina–Zirconia Coatings Using the Solution Precursor Plasma Spray Process, J. Am. Ceram. Soc. 91(2) 359-365 (2008)

[19] H. Hayashi, M. Kanoh, C. J. Quan, H. Inaba, S. Wang, M. Dokiya, H. Tagawa, Thermal expansion of Gd-doped ceria and reduced ceria, Solid State Ionics 132 (2000) 227–233

[20] L.-W. Tai, M. M. Nasrallah, H. U. Anderson, D. M. Sparlin and S. R. Sehlin, Structure and electrical properties of $La1-xSrxCo1-yFeyO3$. Part 1. The system $La_{1-x}Sr_xCo_{0.2}Fe_{0.8}O_3$, Solid State Ionics 76 (1995), 273-283

[21] S. Wang, M. Katsuki, M. Dokiya and T. Hashimoto, High temperature properties of $La0.6Sr0.4Co0.8Fe0.2O3-\delta$ phase structure and electrical conductivity, Solid State Ionics 159 (2003), 71-78

[22] A. Mineshige, J. Izutsu, M. Nakamura, K. Nigaki, J. Abe, M. Kobune, S. Fujii, T. Yazawa, Introduction of A-site deficiency into $La_{0.6}Sr_{0.4}Co_{0.2}Fe_{0.8}O_{3-\delta}$ and its effect on structure and conductivity, Solid State Ionics 176 (2005) 1145– 1149

[23] P. Zeng, R. Ran, Z. Chen, H. Gu, Z. Shao, J.C. Diniz da Costa, S. Liu, Significant effects of sintering temperature on the performance of $La_{0.6}Sr_{0.4}Co_{0.2}Fe_{0.8}O_{3-\delta}$ oxygen selective membranes, Journal of Membrane Science 302 (2007) 171–179

[24] P. Ried, P. Holtappels, A. Wichser, A. Ulrich, and T. Graule, Synthesis and characterization of $La_{0.6}Sr_{0.4}Co_{0.2}Fe_{0.8}O_{3-\delta}$ and $Ba_{0.5}Sr_{0.5}Co_{0.8}Fe_{0.2}O_{3-\delta}$ , Journal of The Electrochemical Society, 155(10) B1029-B1035 (2008)

[25] L.-W. Tai, M. M. Nasrallah, H. U. Anderson, D. M. Sparlin and S. R. Sehlin, Structure and electrical properties of $La1-xSrxCo1-yFeyO3$. Part 1. The system $La_{0.8}Sr_{0.2}Co_{1-y}Fe_yO_3$, Solid State Ionics 76 (1995), 259-271

# EFFECTS OF GEOMETRICAL AND MECHANICAL PROPERTIES OF VARIOUS COMPONENTS ON STRESSES OF THE SEALS IN SOFCS

W.N. Liu, B.J. Koeppel, X. Sun, M.A. Khaleel
Pacific Northwest National Laboratory
902 Battelle Blvd, Richland, WA. 99352

## ABSTRACT

In this paper, numerical modeling was used to understand the effects of the geometry and mechanical properties of various components in SOFCs on the magnitude and distribution of stresses in the stack during operating and cooling processes. The results of these modeling analyses will help stack designers reduce high stresses in the seals of the stack so that structural failures are prevented and high stack mechanical reliability is achieved to meet technical targets. In general, it was found that the load carrying capacity of the cathode contact layer was advantageous for reducing the transmitted loads on the cell perimeter seal under operating environments of SOFCs, but the amount of reduction depends upon the relative stiffness values of the cell, interconnect, porous media, and support structures. Comparison to a frictionless sliding interface, a fully bonded interface resulted in 30-50% less transmitted load through the perimeter seal, with the greater reductions due to stiffer contact/media/interconnect structures. These results demonstrate that the mechanical contribution of the contact layer can be substantial and warrant design consideration.

## INTRODUCTION

As high-efficiency and environment-friendly energy conversion device, solid oxide fuel cells (SOFC) have shown great potential in delivering high performance at reasonable costs [1,2]. Planar SOFCs offer a significant advantage of a compact design along with higher power densities. In the meantime, they require the incorporation of hermetic gas seals for efficient and effective channeling of fuel and oxygen.

Seals are the most critical components in commercializing the planar SOFC technology. They must adequately prevent the leakage of air and fuel, effectively isolate the fuel from the oxidant, and insulate the cell from short circuit. To obtain a reliable SOFC design in the complex operating environments, the stress level in the glass seal materials as well as at the various interfaces must be carefully examined and managed. An understanding of the effects of the geometry and mechanical properties of various components in SOFCs on the level and distribution of stresses in the stack during operating and cooling processes is necessary to quantify the mechanical reliability of the entire cell and seal layers. In this paper, numerical modeling was used to evaluate these thermal-mechanical stresses. The results of these modeling analyses will help stack designers reduce high stresses in the seals of the stack so that structural failures are prevented and high stack mechanical reliability is achieved to meet technical targets. In general, it was found that the load carrying capacity of the cathode contact layer was advantageous for reducing the transmitted loads on the cell perimeter seal under operating environments of SOFCs, but the amount of reduction depends upon the relative stiffness values of the cell, interconnect(IC), porous media, and support structures. Comparison of a fully bonded interface to a frictionless sliding interface resulted in 30-50% less transmitted load through the perimeter seal, with the greater reductions due to stiffer contact/media/interconnect structures.

These results demonstrate that the mechanical contribution of the contact layer can be substantial and warrant design consideration.

GEOMETRY OF ONE CELLS STACK AND FINITE ELEMENT MODEL

The first generation SECA test cell in Pacific Northwest National Laboratory (PNNL) was used for this study [3]. The geometry of one quarter of this one-cell stack is illustrated in Figure 1. Its in-plane dimensions are 157.9mm by 149.5mm, and the total thickness for the 1-cell stack is 5mm. The active area is 50mm by 50mm. The thickness of the anode, electrolyte and cathode are 640μm, 10μm, and 50μm, respectively. A cross-flow configuration was investigated. In the air flow channel, a continuous contact paste layer was applied between the cathode and current collector. The steel-mesh material was adopted to collect the current and flow through the air due to its porous feature. The model was used to evaluate the effects of geometry and mechanical properties on the stack stress state and load path under operating conditions. The stresses and load path for the cell were then evaluated both with and without the effects of contact paste layer densification for comparison.

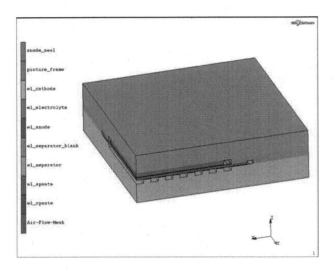

Figure 1. Model geometry based on the SECA Gen1 test cell.

Figure 2 depicts the local structure of the flow channels and positive electrode-electrolyte-negative electrode (PEN) structure. The contact paste was applied beneath the air flow mesh and was continuous thin layer. The thickness of the contact paste is 100μm. The air flow mesh is considered as a uniform and continuous layer. To investigate the effects of the interfacial condition of the contact paste and cathode on the stress and load path in the seal, two extreme conditions were considered, i.e., the contact paste was bonded with cathode, or the contact paste could slide along the cathode surface.

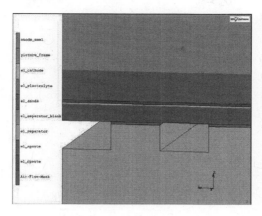

Figure 2 Local cross-section

The FE mesh used in the parametric investigation is shown in Figure 3. Since the SOFC components are considered to be stress free at its assembly temperature of 800°C, different levels of thermal stresses will be generated at the steady state operating temperature due to the mismatch of coefficient of thermal expansion (CTE) of various cell components. The working temperature is assumed to be uniform as 750°C, (1023 K).

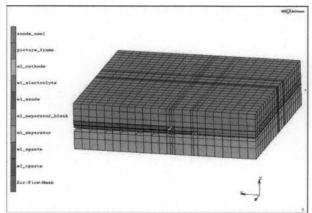

Figure 3 Finite element model used in the thermal-mechanical analysis

## MECHANICAL PROPERTIES OF VARIOUS COMPONENTS IN SOFC STACKS

Except for the seals and the PEN structure in the SOFC stack, all the other materials are assumed to be SS430 in this simulation. All the mechanical properties of the various components in SOFC are temperature dependent. The mechanical properties of a commonly used high temperature ferritic alloy SS430 is used in the current numerical analyses. Figures 4 and 5 depict the temperature dependent Young's modulus and CTE for SS430 [4]. Under the operating temperature of SOFC, creep of SS430 is unavoidable. However, the purpose of this paper is to investigate the effects of geometry and mechanical property of contact paste layer and IC ribs on the stress and load path in the seal, therefore, the creep behavior is not included in the study.

The effects of moduli of the mesh and contact paste on the stress were parametrically investigated. The modulus of the mesh was assigned as 10%, 50%, and 100% of the modulus of steel used as IC. The material of the cathode contact paste material is similar to cathode material but also has a different modulus, as the mechanical properties of the cathode contact paste vary with its porosity. In the parametric study, the modulus of the contact paste material was taken as 10%, 50%, and 100% of the modulus of cathode used in the SOFC stack.

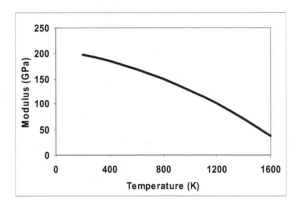

Figure 4 Temperature dependent modulus of SS430

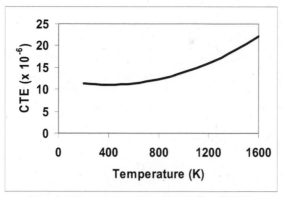

Figure 5 Temperature dependent CTE of SS430

## NUMERICAL RESULTS AND DISCUSSIONS

The typical stress distribution in the PEN seal is illustrated in Figure 6. It can be seen that stress is non-uniform in the seal. Therefore, resultant stress, i.e., force, was used to evaluate the load path.

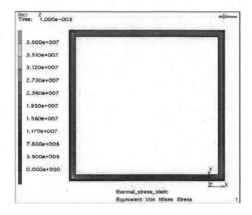

Figure 6 Typical distribution of stress in PEN seal

Figures 7 and 8 include the modeling results of the resultant force in various conditions. In the figures, $E$p and $E$c represent the moduli of the contact paste and cathode, respectively; and $E$m and $E$ic refer to the moduli of the mesh material and steel interconnect, respectively. The modeling results for both sliding and bonded interfaces of the contact paste and cathode were included together in the figures. It may be seen that for every case considered here, the bonded interface between the contact paste and cathode creates lower perimeter seal force than sliding interface between the contact paste and cathode. It means that with a bonded interface between the contact paste and cathode, the stress caused by the mismatch of CTE of PEN structure and steel IC is partially transferred through the mesh. With bonded interface, increased mesh modulus decreases the seal force. On the other hand, with sliding interface, mesh modulus has little effect on seal force. When the interface of the contact paste and the cathode is sliding, the stress/load could not be transferred through the mesh layer by the interfacial shear load; therefore, the modulus of the mesh material barely has any influence on the stresses/load in the PEN seals.

No matter which condition is considered for the interface of the contact paste and cathode, the stiffer contact paste increases seal force slightly. Compared with sliding interface between the contact paste and cathode, however, stiffer contact paste reduces the seal force more effectively with the bonded interface. It was also found that stiffer mesh reduces the seal force more effectively.

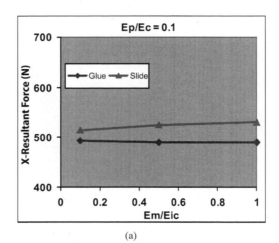

(a)

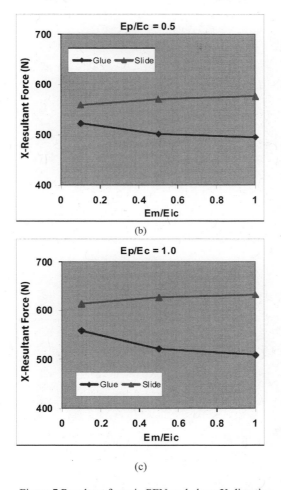

(b)

(c)

Figure 7 Resultant force in PEN seal along X-direction

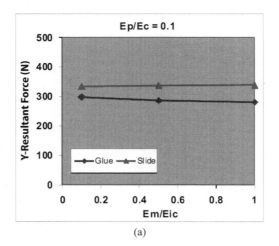

(a)

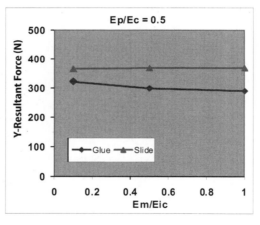

(b)

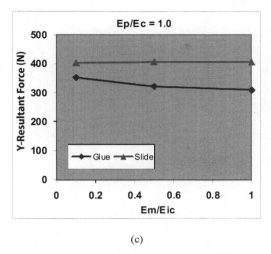

(c)

Figure 8 Resultant force in PEN seal along Y-direction

CONCLUSIONS
The effects of cathode contact paste and IC ribs on the stress and load path in the cell perimeter seal were investigated parametrically. The purpose is to manage the stress and load in the seal for reliable design of SOFC. For the one-cell stack with continuous contact paste layer between the cathode and current collector as well CC mesh in the air channel, the conclusions of the current study can be summarized as follows: In general, it was found that the high load carrying capacity of the cathode contact layer was advantageous for reducing the transmitted loads on the cell perimeter seal under operating environments of SOFCs, but the amount of reduction depends upon the relative stiffness values of the cell, IC, porous media, and support structures. Comparison of a fully bonded interface to a frictionless sliding interface resulted in 30-50% less transmitted load through the perimeter seal, with the greater reductions due to stiffer contact/media/IC structures.

ACKNOWLEDGEMENTS
The Pacific Northwest National Laboratory is operated by Battelle Memorial Institute for the United States Department of Energy under Contract DE-AC06-76RL01830. The work was funded as part of the Solid-State Energy Conversion Alliance (SECA) Core Technology Program by the U.S. Department of Energy's National Energy Technology Laboratory (NETL).

REFERENCES
[1]M. Granovskii, I. Dincer, M.A. Rosen, Exergy analysis of a gas turbine cycle with steam generation for methane conversion within solid oxide fuel cells, *Journal of Fuel Cell Science and Technology*, v 5, n 3, p 031005-1-9, Aug. 2008.

[2]E. Fontell, T. Kivisaari, N. Christiansen, J. B. Hansen, J. Palsson, Conceptual study of a 250 kW planar SOFC system for CHP application, *Journal of Power Sources*, v 131, n 1-2, p 49-56, 14 May 2004.
[3]http://www.netl.doe.gov/publications/proceedings/08/seca/Presentations/1_Jeff_Stevenson_SEC A_StackTest.pdf]
[3]K. I. Johnson, V. N. Korolev, B. J. Koeppel, K. P. Recknagle, M. A. Khaleel, D. Malcolm and Z. Pursell, Finite Element Analysis of Solid Oxide Fuel Cells Using SOFC-MP™ and MSC.Marc/Mentat-FC™, PNNL report 15154, Pacific Northwest National Laboratory, June 2005.

# STABILITY OF MATERIALS IN HIGH TEMPERATURE WATER VAPOR: SOFC APPLICATIONS

E.J. Opila and N.S. Jacobson
NASA Glenn Research Center
Cleveland, OH 44135

## ABSTRACT

Solid oxide fuel cell (SOFC) material systems require long term stability in environments containing high-temperature water vapor. Materials in fuel cell systems may react with high-temperature water vapor to form volatile hydroxides which can degrade cell performance. In this paper, experimental methods to characterize these volatility reactions including mass spectrometry, the transpiration technique, and thermogravimetric analysis are reviewed. Experimentally determined data for chromia, silica, and alumina volatility are presented. In addition, data from the literature for the stability of other materials important in fuel cell systems are reviewed. Finally, methods for predicting material recession due to volatilization reactions are described.

## INTRODUCTION

Solid oxide fuel cells are under development for a variety of power generation applications. These applications require operation for long times, on the order of 20,000 to 100,000 h, at temperatures from 500°C to 1000°C depending on the application. The anode inlet environment is fuel-rich containing large partial pressures of hydrogen or hydrocarbons. As the gas flows across the anode it is depleted in fuel and enriched in water vapor, a product of the cell operation, so that the anode outlet may contain high partial pressures of water vapor. The cathode, or oxidant side of the cell, contains water vapor as part of the air feed or any exhaust that is recycled to the cathode inlet. Total pressures in the cell may be 1 atm for simple SOFC or higher pressures when combined in hybrid systems. In addition to composition and partial pressure, the feed rates of gases play an important role in any gas solid reaction. This paper addresses the interaction of cell materials with the water vapor that is present on both the anode and cathode sides of the cell with the intent of increasing understanding of water vapor-related degradation reactions.

A generic expression for the reaction of high temperature water vapor with a solid phase to form a gaseous metal hydroxide is given in Reaction 1.

$$MO_x + nH_2O(g) + mO_2(g) + qH_2(g) = M O_{(x+n+2m)} H_{(2n+q)}(g) \qquad (1)$$

At the high operating temperatures of SOFCs, the partial pressures of gaseous metal hydroxides depend on the equilibrium thermodynamics of this type of reaction. The partial pressure of the M-O-H(g) species will depend on the partial pressures of the reactant gases: $P(H_2O)^n$, $P(O_2)^m$, and $P(H_2)^q$. In addition to thermodynamics, the transport kinetics of the volatile metal hydroxides in the gas phase are also important towards predicting degradation rates of the cell. Formation of gaseous metal hydroxides can lead to poisoning of other parts of the cell as the volatile species condense and/or react. In addition, the volatilization reactions can lead to material loss of thin layers essential for long term cell operation. Understanding the thermodynamics and kinetics of these reactions allows prediction of degradation rates over a range of operating conditions.

The specific objectives of this paper are to review techniques for characterizing volatilization reactions, to review available thermodynamic data for gaseous metal hydroxides for material systems important in SOFCs, to identify gaps in the available thermodynamic data needed to predict the long term durability of SOFCs, and finally to present the kinetic expressions used to calculate material loss rates from thermodynamic data.

EXPERIMENTAL TECHNIQUES FOR CHARACTERIZING VOLATILIZATION REACTIONS

Several techniques are available to characterize both the thermodynamics and kinetics of gaseous metal hydroxide formation. It is important to first establish the identity of the vapor species, next to determine accurate thermodynamic data for the formation reaction, and finally to understand the kinetics of the volatilization reaction. General aspects of mass spectrometry, the transpiration method, and thermogravimetric analysis pertinent to the characterization of gaseous metal hydroxide formation are discussed in the following paragraphs.

Mass Spectrometry

High temperature mass spectrometry of gaseous metal hydroxides is a technique in which the gas species are formed by reaction with high temperature water vapor, ionized, and separated by mass-to-charge ratio using a quadrupole, magnetic sector, or time-of-flight mass filter. A review of the technique can be found in Reference 1. Accurate thermodynamic data for gaseous metal hydroxides can be determined using a Knudsen Effusion Mass Spectrometer (KEMS) as demonstrated for Si-O-H(g),[2,3] Mn-O-H(g),[4] Pd-O(g).[5,6] In the KEMS technique, the solid under study is contained in an enclosed cell with a very small orifice (<1 mm), low partial pressures of water vapor are introduced to the cell, gas-solid equilibrium is obtained, the product gaseous metal hydroxides leak through the orifice without disturbing the equilibrium and are identified by mass spectrometry. This process is shown in Figure 1. One drawback of the KEMS technique is that the Knudsen Cell must be held in a high vacuum chamber and only very low partial pressures of water vapor ($<10^{-5}$ bar) can be introduced to the Knudsen Cell to establish the equilibrium. As a result the reactant pressures are not representative of processes occurring near one atmosphere as would be found in a SOFC.

Alternatively, a high pressure free jet expansion mass spectrometer can be used to sample gaseous metal hydroxide formation reactions occurring at one atmosphere as shown in Figure 2. The free jet expansion produces an abrupt transition to collisionless flow, thereby preserving the chemical integrity of the process from the reaction chamber to the mass spectrometer. The differentially pumped vacuum chambers allow the beam to be sampled at the low pressures necessary to operate a mass spectrometer. In this system a quadrupole mass spectrometer is used for identification of the vapor species, however, quantitative determination of the partial pressures of the vapor species is extremely difficult. This technique is therefore used for identification of vapor species formed at high pressures with only qualitative determination of the partial pressures of the volatile species as demonstrated for $CrO_2(OH)_2(g)$[8] and $Si(OH)_4(g)$.[9] Mass spectra for $SiO_2$ volatility are shown in Figure 3 for dry oxygen and wet oxygen providing the first direct identification of $Si(OH)_4(g)$. Note that

KEMS of the Si-O-H(g) system[2,3] did not identify $Si(OH)_4(g)$ due to the necessarily low inlet pressures of $H_2O(g)$.

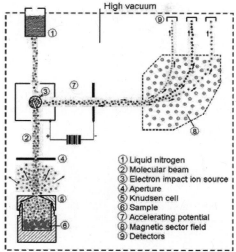

Figure 1. Schematic drawing of the Knudsen Effusion Mass Spectrometry (KEMS) technique.[7]

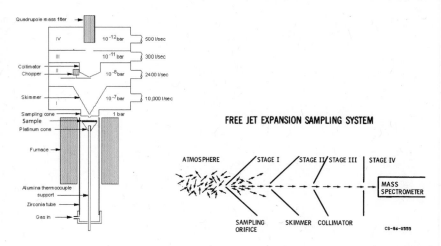

Figure 2. Schematic drawing of a high pressure mass spectrometer and the free jet expansion process.

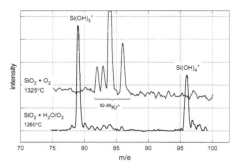

Figure 3. Identification of Si(OH)₄(g) by high pressure mass spectrometry.

The Transpiration Technique

The transpiration technique has been reviewed by Merten and Bell.[10] This is a relatively simple technique for determining thermodynamic data for gas phase species as well as indirectly identifying the gas species at pressures near one atmosphere. In this method, shown schematically in Figure 4, a carrier and/or reactive gas mixture flows over a heated condensed phase sample at rates low enough for equilibrium to be maintained. Any product gases formed from the reaction between the condensed phase and reactant gas are carried downstream to a cooler portion of the gas train where they condense. The condensate is collected and the amount of condensate is determined quantitatively by chemical analysis techniques, such as atomic emission spectroscopy. This technique provides a direct measurement of condensable gas species partial pressure. Thermodynamic data for the reaction can be determined from the temperature dependence of the reaction. The species identity can be determined indirectly from the dependence of the product partial pressure on the reactant gas ($H_2O(g)$ and $O_2(g)$) pressure. This technique is useful at relatively high pressures, 0.1 to 1 atm, and is thus representative of SOFC conditions. This technique has been used successfully to obtain thermochemical data for gaseous metal hydroxide formation from Ni and Co,[11] $SiO_2$,[12,13] CaO,[13,14] and $Cr_2O_3$.[15,16]

Thermogravimetric Analysis (TGA)

TGA is another simple technique for monitoring volatility of materials in reactive gases. A schematic drawing of an apparatus using water vapor as the reactive gas is shown in Figure 5. A coupon of well-defined geometry is suspended from a micro-balance in a tube within a furnace. The flux of volatile species is monitored through weight loss as a function of time and temperature. For a long tube length, a well-defined laminar flow is established over the flat plate coupon allowing determination of the partial pressure of volatile species from the measured flux. It should be emphasized that the weight loss varies with the experimental gas flow rate. For determination of quantitative thermochemical data, the linear gas velocity must be well-defined and accurately known. Flow rates should be chosen so that gas boundary layer interaction with the furnace tube walls is minimized. The calculation of partial pressures from measured fluxes will be discussed in the final section of this paper. Once the partial pressure of volatile species is determined, the dependence of the volatility on reactant gas pressure can be used to indirectly determine the identity of the volatile species. In addition, thermodynamic data can be determined from the dependence of the volatility on the temperature and reactant partial pressures. This technique has been used to obtain thermochemical data for $Al_2O_3$ volatility in water vapor[17] and will be discussed in more detail below.

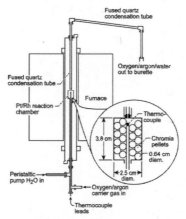

Figure 4. Schematic drawing of a transpiration apparatus used to obtain thermochemical data for gaseous metal hydroxides formed from $Cr_2O_3$.

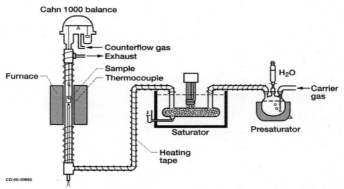

Figure 5. Schematic drawing of a thermogravimetric apparatus used to determine volatility of a sample coupon in high-temperature water vapor.

Thermodynamic Data for Gaseous Metal Hydroxides Important in SOFC

The Cr-O-H(g), Si-O-H(g), and Al-O-H(g) systems have been investigated in our laboratory. Issues relevant to each system as applied to SOFC are discussed in detail below.

Cr-O-H(g)

$Cr_2O_3$ is found in SOFCs as thermally grown oxides on low temperature metallic interconnects or as a component of $La_{1-x}M_xCrO_3$ used for high temperature interconnectors. It is well known that $Cr_2O_3$ forms volatile hydroxides in water vapor leading to chromium poisoning of the cathode. This problem cannot be avoided by using non- $Cr_2O_3$ forming metallic interconnect materials because other slow-growing oxides (e.g. $Al_2O_3$) are not electrical conductors. Ebbinghaus[18] summarized the thermochemical data for the known gaseous chromium hydroxides in 1993. The partial pressure of the Cr-O-H(g) species formed in an environment containing 50% $O_2$/50% $H_2O$ at one atmosphere total pressure are shown as a function of temperature in Figure 6.

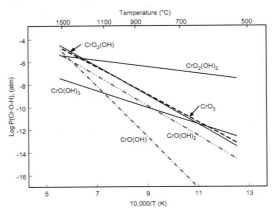

Figure 6. Volatile Cr-O-H(g) species formed in 0.5 atm $H_2O$, 0.5 atm $O_2$ conditions as a function of temperature. Partial pressures were calculated using the data of Ebbinghaus[18].

First, it should be noted that there are many vapor species in this system. It can also be seen that the predominant volatile chromium species are formed by the following reactions:

$$\tfrac{1}{2}\, Cr_2O_3(s) + \tfrac{3}{4}\, O_2(g) = CrO_3(g) \tag{2}$$
$$\tfrac{1}{2}\, Cr_2O_3(s) + H_2O(g) + \tfrac{3}{4}\, O_2(g) = CrO_2(OH)_2(g) \tag{3}$$
$$\tfrac{1}{2}\, Cr_2O_3(s) + \tfrac{1}{2}\, H_2O(g) + \tfrac{1}{2}\, O_2(g) = CrO_2(OH)(g) \tag{4}$$

It is generally accepted that Reaction 3 dominates at conditions of interest for SOFCs although at the time of Ebbinghaus' publication,[18] there was no direct confirmation that $CrO_2(OH)_2(g)$ was the only species forming and there was considerable uncertainty in the thermochemical data used to predict partial pressures of this species. Recent transpiration studies[15,16] have resolved both of these issues as demonstrated in Figures 7 and 8. Figure 7 shows the dependence of $Cr_2O_3$ volatility on both oxygen and water vapor partial pressures at 600°C.[15] The slopes of the lines show the power law dependence for $Cr_2O_3$ volatility, which is in excellent agreement with the coefficients of water and oxygen in Reaction 3. These data demonstrate that under similar conditions, $Cr_2O_3$ volatility can be explained by formation of $CrO_2(OH)_2(g)$ alone. $CrO_3(g)$ begins to make significant contributions to $Cr_2O_3$ volatility at temperatures above 900°C.[15] Figure 8 shows the temperature dependence for Reaction 3 and compares the results of the new studies to other data available in the literature. Because of the good agreement between Opila[15] and Stanislowski,[16] it can be concluded that these values are likely valid. The slope and intercept of these lines correspond to the enthalpy (53.5 kJ/mol) and entropy (-45.6 J/K mol) of Reaction 3 and can therefore be used to predict $Cr_2O_3$ volatility under desired conditions of SOFC operation.

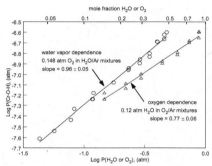

Figure 7. Pressure dependence of $Cr_2O_3$ volatility at 600°C confirming $CrO_2(OH)_2(g)$ as the volatile species.[15]

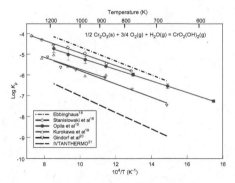

Figure 8. Temperature dependence of the equilibrium constant for Reaction 3 to form $CrO_2(OH)_2(g)$. Data from Opila et al[15] and Stanislowski[16] et al are preferred.

Si-O-H(g)

SiO₂ may be found in SOFCs as a constituent of sealing glasses. Formation of volatile Si-containing species can result in poisoning of the fuel cell anode. $SiO_2$ volatility has been studied in great detail for applications of structural Si-based ceramics in combustion environments.[22] Several volatile silicon hydroxide species have been observed experimentally as detailed in the following reactions.

$$SiO_2(s) + 2\ H_2O(g) = Si(OH)_4(g) \tag{5}$$
$$SiO_2(s) + H_2O(g) = SiO(OH)_2(g) \tag{6}$$
$$SiO_2(s) + \tfrac{1}{2}\ H_2O(g) = SiO(OH)(g) + \tfrac{1}{4}\ O_2(g) \tag{7}$$

$Si(OH)_4(g)$ is likely to be the predominant species under SOFC operating conditions and has been observed by high pressure mass spectrometry (Figure 3),[9] transpiration[12,13] and TGA.[23] $SiO(OH)_2(g)$ and $SiO(OH)(g)$ have been observed at higher temperatures and/or lower water vapor partial pressures by transpiration[12] and Knudsen Effusion Mass Spectrometry[2,3] and are unlikely to be important for SOFC applications. Pressure- and temperature-dependent results for SiO₂ volatility results from transpiration experiments are shown in Figures 9 and 10 respectively. Figure 9 show that SiO₂ volatility follows a square dependence on the water vapor partial pressure consistent with $Si(OH)_4(g)$ formation. In addition, Figure 10 shows that the temperature dependence for $Si(OH)_4(g)$ formation is relatively low. The thermochemical data for $Si(OH)_4(g)$ are now well established[12,13] enabling accurate prediction of SiO₂ volatility rates in high temperature water vapor.

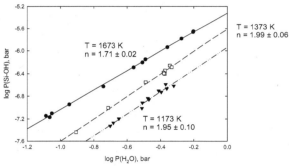

Figure 9. Pressure dependence of SiO₂ volatility consistent with $Si(OH)_4(g)$ formation at 1173 and 1373K.[12] Contributions of other Si-O-H(g) species are significant at 1673K, as indicated by slope < 2.

Al-O-H(g)

Al₂O₃ may be found in SOFCs as a constituent of sealing glasses, insulators, and/or gas delivery components. Al₂O₃ volatility in high-temperature water vapor has been studied in some detail by both transpiration experiments[13] and TGA[17]. Based on the pressure dependence of the volatility, aluminum hydroxide formation has been attributed to the following reaction:

$$1/2\ Al_2O_3\ (s) + 3/2\ H_2O(g) = Al(OH)_3(g) \tag{8}$$

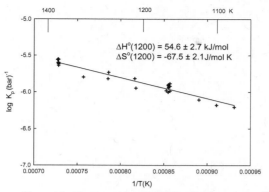

Figure 10. Temperature dependence of SiO₂ volatility.[12]

Good agreement is found between the experimental work and estimated data found in several data bases as shown in Figure 11. The partial pressures of this species are expected to be low at the operating temperatures of SOFCs, nevertheless this work is included here to provide an example for the determination of thermodynamic data through the measurement of weight loss by TGA.[17] It is important to note that the weight loss depends on the flow rate of the gas stream and the volatility is limited by transport of the gaseous products away from the surface through a gas boundary layer. In the TGA experiments the gas boundary layer is laminar, but under some conditions the flow can be less well defined or turbulent. Kinetic modeling of gas boundary layer limited volatilization will be described in more detail in the final section of this paper.

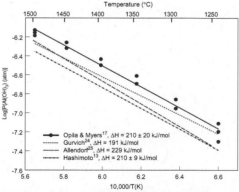

Figure 11. Temperature dependence of Al(OH)₃(g) formation by Reaction 8 showing good agreement between transpiration results,[13] TGA,[17] and literature results.[24,25]

The relative volatility of $Cr_2O_3$, $SiO_2$, and $Al_2O_3$ are compared in Figure 12. It can be seen that $Cr_2O_3$ is much more volatile than the other oxides.

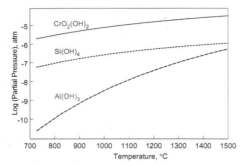

Figure 12. A comparison of the partial pressures of predominant gaseous metal hydroxides formed from $Cr_2O_3$, $SiO_2$, and $Al_2O_3$ in an 0.5 atm $O_2$(g)/0.5 atm $H_2O$(g) environment as a function of temperature.

SOFC Materials Systems with Known Thermochemical Data for Gaseous Metal Hydroxides

In addition to the data obtained in our laboratory, a significant amount of good experimental data for gaseous metal hydroxides significant in SOFC systems is available. For example, Ni anode reactions with high temperature water vapor can be predicted based on the transpiration data of Belton and Jordan[11] as compiled in the Fact database.[26] $Ni(OH)_2$(g) is predicted to be the dominant volatile species. Data for $Ni(OH)$(g), $NiH$(g) are found in the SGTE data base[27] but their quality is uncertain. Based on the results of Belton and Jordon,[11] both $Ni(OH)$(g) and $NiH$(g) should be negligible. The SGTE data[27] over-predicts the stability of $Ni(OH)$(g) and $Ni(H)$(g) may therefore be unreliable. The paper by Belton and Jordon[11] also contains reliable data for $Co(OH)_2$(g) formation which is important for interconnect alloys and coatings containing cobalt.

SrO and CaO gaseous hydroxide formation have been studied using atomic absorption spectroscopy measurements with hydrogen flames.[28,29] The stability of these oxides is important in SOFC systems since SrO is incorporated in $La_{1-x}Sr_xMnO_{3-\delta}$ cathodes and CaO is incorporated in high-temperature $LaCrO_3$ interconnects. $Sr(OH)_2$(g) and $Ca(OH)_2$(g) are predicted to be the primary volatile species in cathode conditions. There is widespread agreement across the databases on thermodynamic data for these species.[30] It should be noted that $Ca(OH)_2$(g) partial pressures are about an order of magnitude lower than $Sr(OH)_2$(g) for the reaction of the simple oxides with high temperature water vapor. It is important to consider the reduced activity of these oxides in the complex oxides when predicting the partial pressures of these volatile species under SOFC operating conditions. For example, Peck et al[31] have determined the activities of CaO, $Cr_2O_3$, and $La_2O_3$ in $La_{1-x}Ca_xCrO_{3-\delta}$, x=0-0.21 at 2000K and found CaO activity to be ideal while the $Cr_2O_3$ activity is reduced by as much as four orders of magnitude. The reduced activity of $Cr_2O_3$ in $La_{1-x}Ca_xCrO_{3-\delta}$ will likewise result in four orders of magnitude reduction in $Cr_2O_3$ volatility from this material relative to pure $Cr_2O_3$.

SOFC Materials Systems with Unknown Thermochemical Data for Gaseous Metal Hydroxides

Literature was also reviewed for thermochemical data of gaseous metal hydroxides of other components of SOFC material systems. Various volatile species are predicted to form from $La_2O_3$, MnO, Pd, and Pt in high temperature environments, however, the data for gaseous metal hydroxides from these systems are either missing, incomplete, or estimated as summarized below.

The stability of $La_2O_3$ is important to the long term durability of SOFC since it is a major constituent of both $La_{1-x}Sr_xMnO_{3-\delta}$ cathodes and $LaCrO_3$ interconnects. The reaction of high temperature water vapor with $La_2O_3$ has apparently not been studied. Heyrman et al[32] have thoroughly reviewed the vaporization of this oxide and $LaO(g)$ is most likely the predominant vapor in the absence of water vapor. Jackson[33] has estimated thermochemical data for mono- and di-hydroxides of all elements, including $La(OH)(g)$ and $La(OH)_2(g)$. Cubiciotti[34] reports data for $La(OH)_3(g)$ with uncertain origin. Thus the data for gaseous metal hydroxides of $La_2O_3$ are of low reliability. Using these data, all of uncertain origin, the stability of $La_2O_3$ in high temperature water vapor is predicted to be quite high with $La(OH)_3(g)$ predicted to be the most dominant but still negligible: $P=10^{-16}$ bar at 900°C. However, experimental work to confirm these estimates is needed.

The stability of MnO in SOFC is also critical since it is a component of $La_{1-x}Sr_xMnO_{3-\delta}$ cathodes, a component of $(Mn,Cr)_3O_4$ spinels formed on low temperature metallic interconnects such as Crofer 22 APU, and a component of $(Co,Mn)_3O_4$ spinel coatings applied to metallic interconnects to reduce $CrO_2(OH)_2(g)$ formation. The reaction of MnO with high temperature water vapor at atmospheric pressure has apparently not been well studied either. Using data from available thermochemical data bases,[26,27] $MnOH(g)$, $MnH(g)$, and $Mn(g)$ are predicted to form at low partial pressures that would result in minimal problems for long-term operation of SOFC. Hildenbrand & Lau have conducted KEMS studies of MnO at low partial pressures of water vapor[4] and observed $Mn(OH)_2(g)$ and $MnO(OH)(g)$ in addition to the previously reported species. Using the data of Hildenbrand and Lau,[4] thermochemical data for these species have been estimated[35] and the formation of these species are predicted to be more important than $MnOH(g)$ for SOFC applications. Hildenbrand[4] also points out that at high partial pressures of water vapor the species $Mn(OH)_3(g)$ and $MnO(OH)_2(g)$ are likely to become important, however, no data are available for these species. This is consistent with pressure-dependent observations from the Si-O-H(g) system as previously noted. Experimentally observed volatility rates of Mn-containing alloys and coatings[16] also indicate that Mn-O-H(g) formation may be a long-term durability issue for SOFCs. Thus the characterization of MnO stability in high temperature water vapor at high partial pressures is still needed.

Pd and Pt are used as current collectors in SOFCs. The stability of these metals in high temperature water vapor is also a concern. Despite the use of these materials as catalysts in many high temperature systems, data for gaseous metal hydroxides are unknown. Hildenbrand et al[6] report $Pd(g)$ and $PdO(g)$ form in relatively low partial pressures as determined by KEMS in the absence of water vapor. Estimated data for $Pd(OH)(g)$ and $Pd(OH)_2(g)$ are found in the report by Jackson,[33] however, these species have not been experimentally observed. Nevertheless, using the Jackson data[33] it appears that formation of $Pd(OH)(g)$ could be a possible concern for long term durability of thin Pd layers in SOFC conditions. Experimental work is needed to confirm the formation of Pd-O-H(g) species.

Similarly, no known experimental data exist for Pt-O-H(g) species. Jackson again reports some estimated data for $Pt(OH)(g)$ and $Pt(OH)_2(g)$, however, it is well known that $PtO_2(g)$ is a very stable gas species so that volatilization of the oxide vapor is of great concern. Unpublished work[36] demonstrates that weight loss of Pt in Ar/H$_2$O gas mixtures is undetectable at 1200°C whereas Pt weight loss in both Ar/O$_2$ gas mixtures and H$_2$O/O$_2$ is equal and substantial at 1200°C suggesting that O$_2$ is the only reacting gas. While experimental Pt-O-H(g) data are unavailable, the formation of PtO$_2$(g) is expected to dominate the durability of Pt films in SOFC cathode conditions. Data for PtO$_2$(g) are currently available in thermochemical databases.[26,27]

Table I summarizes predicted partial pressures for all gaseous metal hydroxides considered in this paper as a means for making some conclusions about which components of SOFC are particularly susceptible to volatilization. Using FactSage free energy minimization software,[26] partial pressures of volatile species were calculated at two temperatures, 500 and 1000°C, and for gas mixtures of 50% H$_2$/50% H$_2$O and 50% O$_2$/50% H$_2$O at 1 bar total pressure to show the range of expected volatility in both model anode and cathode conditions. The table shows the dominant volatile species for each material as well as the reliability of the calculated partial pressure. It should be noted that when the gaseous metal hydroxide species and partial pressure do not vary between "anode" and "cathode" conditions, the formation reaction is independent of oxygen partial pressure and depends only on water vapor partial pressure as is the case for SiO$_2$ volatility (Reaction 5). Cr$_2$O$_3$ is much more stable in H$_2$/H$_2$O environments than O$_2$/H$_2$O environments since the three major volatile species, CrO$_3$(g), CrO$_2$(OH)$_2$(g), and CrO$_2$(OH)(g) all require oxygen in their formation reactions (Equations 2-4). Alternatively, the dominant gaseous metal hydroxide species can be the same in H$_2$/H$_2$O and O$_2$/H$_2$O environments, and the partial pressures can vary as is the case for Ni and CoO, since the stable solid product (Ni, NiO and Co, CoO, Co$_3$O$_4$) varies with the environment.

## PREDICTION OF VOLATILITY RATES

Up to this point the thermodynamics of gaseous metal hydroxide formation have been discussed. However, volatility reactions in SOFCs will be limited by transport of the equilibrium gaseous metal hydroxides through gas boundary layers. A schematic of a laminar gas boundary layer limited volatilization of SiO$_2$ for a flat plate geometry of characteristic length L is shown in Figure 13. At the solid surface, the gas velocity is zero. At the edge of the gas boundary layer, the gas velocity reaches the free stream gas velocity. The thickness of the boundary layer, δ, depends on the free stream velocity, so that as the velocity increases the boundary layer thickness decreases allowing more rapid transport of gaseous metal hydroxides away from the solid surface. This type of geometry is present in TGA experiments for flat plate coupons.[17] By understanding the flow characteristics, the equilibrium partial pressure of the gaseous metal hydroxide formed at the solid surface can be determined from the weight loss, or flux of the volatile species.

An expression for the flux of volatile species for the geometry shown in Figure 13 is given as follows:[37]

$$J = 0.664 \left( \frac{\rho' v L}{\eta} \right)^{1/2} \left( \frac{\eta}{\rho' D} \right)^{1/3} \frac{D\rho}{L} \tag{9}$$

Here, J is the flux or weight loss rate (mg/cm$^2$ hr), the term in the first parenthesis is the Reynolds number, the term in the second parenthesis is the Schmidt number, ρ' is the gas density of the boundary layer, v is the linear gas velocity, L is a characteristic dimension of the component, η is the gas viscosity, D is the interdiffusion coefficient of the volatile species in the gas boundary layer, and $\rho$ is the gas density of the volatile species.

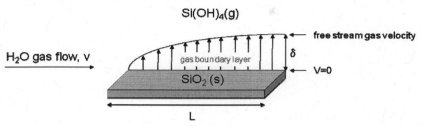

Figure 13. Schematic drawing of SiO₂ volatility limited by transport through a laminar gas boundary layer.

While Equation 9 is accurate for a flat plate geometry with laminar flow and useful for deriving equilibrium partial pressures of gaseous metal hydroxides from TGA experiments with coupons, deviations may exist for real SOFC systems. First, any deviations from laminar flow present in the actual cell environment should be considered. The flat plate geometry may be a good representation for planar cells, but an expression for mass transport from a tube wall in laminar flow may be a more realistic geometry for tubular SOFC components.[38] In addition, the active surface of the porous anode and cathode should be accounted for using a more complete model which accounts for transport in both the pores as well as the boundary layer in the fuel cell tube. This geometry has been considered before for annular pores in ceramic matrix composites.[39]

SUMMARY AND CONCLUSIONS

Methods for characterizing the identity and partial pressure of gaseous metal hydroxides including mass spectrometry, transpiration, and thermogravimetric analysis have been described. Available thermochemical data for gaseous metal hydroxides of materials important in SOFC systems have been summarized. These materials include Cr₂O₃, SiO₂, Al₂O₃, Ni, CoO, SrO, CaO, La₂O₃, MnO, Pd, and Pt. While some thermochemical data are well established, additional thermochemical data are needed for La₂O₃, MnO, and Pd hydroxide vapor species. Variations in equilibrium partial pressure of gaseous metal hydroxides with temperature and environment have been presented. Finally, volatilization kinetics limited by transport through a laminar gas boundary layer was discussed.

Table I. Dominant volatile species and their calculated partial pressures (bar) in model cathode (1 bar, 50% $O_2$/50% $H_2O$) and anode (1 bar, 50% $H_2$/50% $H_2O$) conditions.

| reactant solid | data source | prediction reliability | cathode, $O_2 + H_2O$ | | anode, $H_2 + H_2O$ | | comments |
|---|---|---|---|---|---|---|---|
| | | | 500°C | 1000°C | 500°C | 1000°C | |
| $Cr_2O_3$ | Fact[26] + Opila[15] | high | $CrO_2(OH)_2$ 9E-07 | $CrO_2(OH)_2$ 2E-05 | $Cr(OH)_2$ 2E-19 | $CrO(OH)$ 3E-10 | $H_2$: $Cr(OH)_2(g)$, 500-900°C |
| $SiO_2$ | Fact[26] + Jacobson[12] | high | $Si(OH)_4$ 7E-08 | $Si(OH)_4$ 1E-06 | $Si(OH)_4$ 7E-08 | $Si(OH)_4$ 1E-06 | |
| $Al_2O_3$ | Fact[26] + Allendorf[25] | high | $Al(OH)_3$ 9E-16 | $Al(OH)_3$ 1E-09 | $Al(OH)_3$ 9E-16 | $Al(OH)_3$ 1E-09 | |
| $Ni$ | Fact[26] | high | $Ni(OH)_2$ 6E-13 | $Ni(OH)_2$ 4E-07 | $Ni(OH)_2$ 1E-15 | $Ni(OH)_2$ 2E-09 | solid products: Ni in $H_2$, NiO in $O_2$ |
| $CoO$ | SGTE[27] | unknown | $Co(OH)_2$ 2E-14 | $Co(OH)_2$ 3E-07 | $Co(OH)_2$ 1E-12 | $Co(OH)_2$ 5E-07 | solid products: $Co_3O_4$ in $O_2$ (500-900°C), CoO in $O_2$ (1000°C), CoO + Co in $H_2$ |
| $SrO$ | Fact[26] | high | $Sr(OH)_2$ 3E-15 | $Sr(OH)_2$ 1E-07 | $Sr(OH)_2$ 3E-15 | $Sr(OH)_2$ 1E-07 | condensed phase products: $Sr(OH)_2$ + SrO (500-600°C), SrO all other temps |
| $CaO$ | Fact[26] | high | $Ca(OH)_2$ 6E-16 | $Ca(OH)_2$ 7E-09 | $Ca(OH)_2$ 6E-16 | $Ca(OH)_2$ 7E-09 | |
| $La_2O_3$ | Fact[26] | low | $La(OH)_3$ 1E-23 | $La(OH)_3$ 1E-14 | $La(OH)_2$ 4E-22 | $La(OH)_2$ 6E-12 | |
| $MnO$ | SGTE[27] + Hildenbrand[4] | low | $Mn(OH)_2$ 6E-18 | $Mn(OH)_2$ 1E-09 | $Mn(OH)_2$ 9E-15 | $MnH$ 4E-08 | $H_2$: $Mn(OH)_2(g)$ (500-700°C), MnH(g) 800-1000°C; solid products: MnO in $H_2$, $Mn_2O_3$ in $O_2$, $Mn_3O_4$ in $O_2$ at 1000°C |
| $Pd$ | Fact[26] | low | $Pd(OH)$ 9E-12 | $Pd(OH)$ 2E-05 | $Pd(OH)$ 1E-17 | $Pd(OH)$ 6E-09 | solid products: Pd + PdO in $O_2$ (500-800°C), Pd in $O_2$ (900-1000°C) and $H_2$ all temps |
| $Pt$ | Fact[26] | medium | $PtO_2$ 4E-12 | $PtO_2$ 1E-07 | $Pt$ <1e-25 | $Pt$ 4E-16 | no experimental gaseous hydroxide data |

ACKNOWLEDGMENTS
The thermodynamic data evaluation contained herein was made possible, in part, through funding from Rolls-Royce Fuel Cell Systems, Inc. under NASA Space Act Agreement SAA3-1031. This material is based upon work supported by the Department of Energy National Energy Technology Laboratory under Award Number DE-FE0000303."

Disclaimer: "This report was prepared as an account of work sponsored by an agency of the United States Government. Neither the United States Government nor any agency thereof, nor any of their employees, makes any warranty, express or implied, or assumes any legal liability or responsibility for the accuracy, completeness, or usefulness of any information, apparatus, product, or process disclosed, or represents that its use would not infringe privately owned rights. Reference herein to any specific commercial product, process, or service by trade name, trademark, manufacturer, or otherwise does not necessarily constitute or imply its endorsement, recommendation, or favoring by the United States Government or any agency thereof. The views and opinions of authors expressed herein do not necessarily state or reflect those of the United States Government or any agency thereof."

REFERENCES
1. R.T. Grimley, "Mass Spectrometry" pp. 195-243 in *The Characterization of High Temperature Vapors*, ed. J.L. Margrave, John Wiley & Sons, NY, 1967.
2. D. L. Hildenbrand, K. H. Lau, "Thermochemistry of gaseous SiO(OH), SiO(OH)$_2$, and SiO$_2$," J. Chem. Phys., 101 [7] 6076–79 (1994).
3. D. L. Hildenbrand, K. H. Lau, "Comment on 'Thermochemistry of gaseous SiO(OH), SiO(OH)$_2$, and SiO$_2$'," J. Chem. Phys., 108 [15] 6535 (1998).
4. D.L. Hildenbrand, K.H. Lau, "Thermochemistry of Gaseous Manganese Oxides and Hydroxides," J. Chem. Phys. 100 [11] 8377-8380 (1994).
5. J.H. Norman, H.G. Staley, W.E. Bell, "Mass Spectrometric Knudsen Cell Measurements of the Vapor Pressure of Palladium and the Partial Pressure of Palladium Oxide," J. Phys. Chem. 69 [4] 1373-1376 (1965).
6. D.L. Hildenbrand, K.H. Lau, "Dissociation energy of the PdO Molecule," Chem. Phys. Lett. 319, 95-98 (2000).
7. http://www.fz-juelich.de/ief/ief-2/datapool/page/227/KEMSschematisch.jpg
8. G.C. Fryburg, R.A. Miller, F.J. Kohl, C.A. Stearns, J. Electrochem. Soc., "Volatile Products in the Corrosion of Cr, Mo, Ti, and Four Superalloys Exposed to O$_2$ Containing H$_2$O and Gaseous NaCl," 124 [11] 1738-1743 (1977).
9. E.J. Opila, D.S. Fox, N.S. Jacobson, "Mass Spectrometric Identification of Si-O-H(g) Species from the Reaction of Silica with Water Vapor at Atmospheric Pressure," J. Am. Ceram. Soc. 80 [4] 1009-1012 (1997).
10. U. Merten and W.E. Bell, "The Transpiration Method," pp. 91-114 in *The Characterization of High Temperature Vapors*, ed. J.L. Margrave, John Wiley & Sons, NY, 1967.
11. G.R. Belton, A.S. Jordan, "The Gaseous Hydroxides of Cobalt and Nickel," J. Phys. Chem. 71 [12] 4114-4120 (1967).

12. N.S. Jacobson, E.J. Opila, D. Myers, E. Copland, "Thermodynamics of Gas Phase Species in the Si-O-H System," J. Chem. Thermo. 37, 1130-37 (2005).
13. A. Hashimoto, "The effect of $H_2O$ gas on volatilities of planet-forming major elements: I. Experimental determination of the thermodynamic properties of Ca-, Al-, and Si-hydroxide gas molecules and its application the solar nebula," Geochim. Cosmo. Acta 56, 511-532 (1992).
14. K. Matsumoto, T. Sata, "A Study of the Calcium Oxide-Water Vapor System by Means of the Transpiration Method," Bull. Chem. Soc. Jpn. 54 [3] 674-677 (1981).
15. E.J. Opila, D.L. Myers, N.S. Jacobson, I.B. Nielsen, D.F. Johnson, J.K. Olminsky, M.D. Allendorf, "Theoretical and Experimental Investigation of the Thermochemistry of $CrO_2(OH)_2(g)$," J. Phys. Chem. A. 111, 1971-1980 (2007).
16. M. Stanislowski, E. Wessel, K. Hilpert, T. Markus, L. Singheiser, "Chromium Vaporization from High-Temperature Alloys," J. Electrochem. Soc. 154 [4] A295-A306 (2007).
17. E.J. Opila and D.L. Myers, "Alumina Volatility in Water Vapor at Elevated Temperatures," J. Am. Ceram. Soc. 87 [9] 1701-1705 (2004).
18. B.B. Ebbinghaus, "Thermodynamics of Gas Phase Chromium Species: The Chromium Oxides, the Chromium Oxyhydroxides, and Volatility Calculations in Waste Incineration Processes," Combust. Flame 93, 119 (1993).
19. H. Kurokawa, C.P. Jacobson, L.C. DeJonghe, S.J. Visco, "Chromium vaporization of uncoated and of coated iron chromium alloys at 1073K," unpublished.
20. C. Gindorf, L. Singheiser, K. Hilpert, "Vaporisation of chromia in humid air," J. Phys. Chem. Solids 66 [2-4] 384-387 (2005)
21. IVTANTHERMO for Windows, version 3.0, 1992-2003.
22. E.J. Opila, J.L. Smialek, R.C. Robinson, D.S. Fox, N.S. Jacobson, "SiC Recession due to $SiO_2$ Scale Volatility under Combustion Conditions. Part II: Thermodynamics and Gaseous Diffusion Model," J. Am. Ceram. Soc. 82 [7] 1826-34 (1999).
23. E.J. Opila, R.E. Hann, "Paralinear Oxidation of CVD SiC in Water Vapor," J. Am. Ceram. Soc., 80 [1] 197-205 (1997).
24. L.V. Gurvich, I.V. Veyts, C.B. Alcock, Thermodynamic Properties of Individual Substances, Begell House, NY, NY, 1996.
25. M.D. Allendorf, C.F. Melius, B. Cosic, A. Fontijn, "BAC-G2 Predictions of Thermochemistry for Gas-Phase Aluminum Compounds," J. Phys. Chem. A [106] 2629-2640 (2002).
26. C.W. Bale, P. Chartrand, S.A. Decterov, G. Eriksson, K. Hack, R. Ben Mahfoud, J. Melançon, A.D. Pelton and S. Petersen, "FactSage Thermochemical Software and Databases", Calphad Journal, 62, 189-228 (2002).
27. A. Dinsdale, "SGTE Data for Pure Elements," Calphad [15] 317-425 (1991).
28. D.H. Cotton, D.R. Jenkins, "Dissociation Energies of Gaseous Alkaline Earth Hydroxides," Trans. Faraday Soc. 64, 2988-2997 (1968).
29. T.M. Sugden, K. Schofield, "Heats of Dissociation of Gaseous Alkali Earth Dihydroxides," Trans. Faraday Soc. 62, 566-575 (1966).
30. M. W. Chase Jr., C. A. Davies, J. R. Downey Jr., D. J. Frurip, R. A. McDonald, A. N. Syverud (Eds.), JANAF Thermochemical Tables, 3rd ed. American Chemical Society and American Physical Society, New York, 1985.
31. D.-H. Peck, M. Miller, K. Hilpert, "Vaporization and thermodynamics of $La_{1-x}Ca_xCrO_{3-\delta}$ investigated by Knudsen effusion mass spectrometry," Solid State Ionics 143, 391-400 (2001).

32. M. Heyrman, C. Chatillon, and A. Pisch, "Congruent vaporization properties as a tool for critical assessment of thermodynamic data: The case of gaseous molecules in the La-O and Y-O systems," Computer Coupling of Phase Diagrams and Thermochemistry 28, 49-63 (2004).

33. D.D. Jackson, "Thermodynamics of the Gaseous Hydroxides," UCRL-51137, Lawrence Livermore National Labs, Livermore, CA, 1971.

34. D. Cubbiciotti, "Vapor Transport of Fission Products under Nuclear Accident Conditions," J. Nucl. Mater. 154, 53-61 (1988).

35. N. Jacobson, D. Myers, E. Opila, E. Copland, "Interactions of water vapor with oxides at elevated temperatures," J. Phys. Chem. Solids 66, 471-478 (2005).

36. E.J. Opila, unpublished.

37. D.R. Gaskell, An Introduction to Transport Phenomena in Materials Engineering, McMillan Publishing Co., NY, NY, 1992.

38. C.J. Geankoplis, pp. 261 in Mass Transport Phenomena, Ohio State Univ. Bookstore, 1978.

39. N.S. Jacobson, G.N. Morscher, D.R. Bryant, R.E. Tressler, "High-Temperature Oxidation of Boron Nitride: II, Boron Nitride Layers in Composites," J. Am. Ceram. Soc. 82 [6] 1473-1482 (1999).

OXYGEN DIFFUSION IN $Bi_2M_4O_9$ (M = Al, Ga, Fe) SYSTEMS AND THE EFFECT OF Sr DOPING IN $Bi_{2-2x}Sr_{2x}M_4O_{9-x}$ STUDIED BY ISOTOPE EXCHANGE EXPERIMENTS AND IR ABSORPTION

T. Debnath[1] , C. H. Rüscher[1*], Th. M. Gesing[1**], P. Fielitz[2], S. Ohmann[2], G. Borchardt[2]

[1]Institut für Mineralogie, Leibniz Universität Hannover, Callinstr. 3, 30167 Hannover, Germany.
[2]Institut für Metallurgie, Arbeitsgruppe Thermochemie und Mikrokinetik, TU Clausthal, Robert-Koch-Straße 42, 38676 Clausthal-Zellerfeld, Germany.
**present adr.: FB05, Kristallographie, Universität Bremen, 28359 Bremen, Germany.
*Corresponding author

ABSTRACT
        Compounds $Bi_{2-2x}Sr_{2x}M_4O_{9-x}$ (M = Al, Ga, Fe) possessing mullite type structure could be interesting candidates for the use as electrolytes in SOFCs at intermediate temperatures (600-800°C). A central role for enhanced oxygen diffusion could be related to the vacancy formation on the site of bridging oxygen of $M_2O_7$ units related to the effect of Sr doping. Polycrystalline samples of average crystal sizes between 20 and 500 nm were synthesized using precursor gels from corresponding nitrate-glycerin solutions. The systems were characterized using XRD, infrared absorption and SEM/EDX techniques. Oxygen tracer diffusion coefficients could be estimated from isotope exchange experiments at 800°C using well separated infrared absorption peaks for the bridging oxygen. The estimated oxygen diffusion coefficients turns out to be in the range $10^{-16}$ to $10^{-18}$ cm$^2$/s for the basic systems and could not be improved with a nominal increase in Sr content, too. This could be explained by the fact that Sr could not be substituted as to form a thermodynamically stable single phase but shows a clear phase separation in acid leaching experiments.

INTRODUCTION
        The family of compounds $Bi_2M_4O_9$ with M = Ga, Al, Fe and their series of solid solution can be described within the family tree of mullite type crystal structure [1], which was not been realized before, although the structure was well known [e.g. 2, 3]. It has been suggested that these Bi-oxide based ceramics could be important candidates for the use as electrolytes of solid oxide fuel cells (SOFCs), gas separation membranes and oxygen sensors due to their high thermal stability. In particular the work of Zha et al. [4] attracted much attention reporting conductivity values of the order of 0.28 S/cm for Sr doped $(Bi_{2-2x}Sr_{2x}Al_4O_{9-x})$ ceramics with x about 0.1 compared to 0.01 S/cm for $Bi_2Al_4O_9$. The suggested high oxygen diffusivity for bismuth aluminate ($Bi_2Al_4O_9$), bismuth gallate ($Bi_2Ga_4O_9$) and in particular strontium doped bismuth aluminate ($Bi_{2-2x}Sr_{2x}Al_4O_{9-x}$) was related to the formation of oxygen vacancy in the $O_c$ site (2d crystallographic site of space group Pbam) [4-7], Here $O_c$ denotes the common oxygen of the double tetrahedra $M_2O_7$ ($\approx O_3M-O_c-MO_3$). In the crystal structure of $Bi_2M_4O_9$, $M^{3+}$ ions are both octahedrally ($MO_6$) and tetrahedrally ($MO_4$) coordinated with oxygen atoms. $MO_6$ octahedra are edge shared forming columns along c-axis (Fig. 1). The octahedra are linked by dimer ($M_2O_7$) of $MO_4$ tetrahedra, forming five-membered rings of two octahedra ($MO_6$) and three tetrahedra ($MO_4$). As a result, a channel like structure is formed parallel to the octahedral chains, where $Bi^{3+}$ ions are located. The $BiO_4$ groups are highly asymmetric and alternate with the planes of $M_2O_7$ units. The unoccupied oxygen sites, which are created due to the stereochemical activity of Bi: $6s^2$ lone pairs, alternate with the $M_2O_7$ dimer along c-axis. Abrahams et al. [5] prepared polycrystalline $Bi_2Al_4O_9$ by conventional solid state synthesis method using equimolar amounts of corresponding oxides. These authors, reported about 6% $\alpha$-$Al_2O_3$ as an impurity phase.

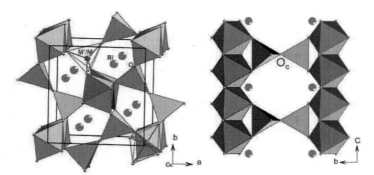

Fig. 1 Structural model of mullite-type $Bi_2M_4O_9$ phase.

Larose et al. [8] reported the presence of $Bi_{24}Al_2O_{39}$ phase during preparation of $Bi_2Al_4O_9$. Zha et al [4] used a combustion method reporting the synthesis of solid solution type series for ($Bi_{2-2x}Sr_{2x}Al_4O_{9-x}$) with x up to about 0.2. The formation of impurity phases could probably also not be avoided in such type of synthesis routes. Moreover, evaporation loss of $Bi_2O_3$ during calcination at high temperature is another factor. Such problems could be eliminated through the formation of organic precursors in the synthesis route. Dissolving salts of the constituent cations into a proper organic solvent generally tends to decrease calcination temperature to reach the desired product. Therefore, organic precursor method were used and it could be shown that single phase multi-cation oxide powders at relatively low temperatures (800°C) was obtained as solid solution type series for the systems $Bi_2(Ga_xAl_{1-x})_4O_9$, $Bi_2(Fe_xAl_{1-x})_4O_9$ and $Bi_2(Fe_xGa_{1-x})_4O_9$ [9, 10]. Moreover, using $^{18}O/^{16}O$ exchange experiments at 800°C evaluated by infrared absorption spectra could be used to estimate tracer diffusion coefficients of about $1.1*10^{-18}$ and $4*10^{-17}$ cm$^2$/s for $Bi_2Al_4O_9$ and $Bi_2Ga_4O_9$, respectively [9, 11] ruling out extraordinary high oxygen diffusivity in these systems. Indications that Sr could not be substituted in a stable form which segregates into an extra phase during heating above 800°C were also reported [12]. Within these investigations, the interpretation of the highest wavenumber infrared absorption peak as assigned to the $M-O_c-M$ bond becomes crucial. Therefore, in the present contribution the role of the $O_c$ oxygen in the infrared absorption spectra was further investigated together with $^{18}O/^{16}O$ exchange experiments conducted for the basic systems ($Bi_2M_4O_9$) with M = Al, Ga, Fe and for nominal compositions $Bi_{2-2x}Sr_{2x}M_4O_{9-x}$ with M = Al, Ga. Acid leaching experiments were carried out, which showed that Sr could not be doped to appropriate amount into the mullite type phase.

EXPERIMENTAL

The following chemicals (all Sigma Aldrich) were used: $Bi(NO_3)_3*5H_2O$, $Al(NO_3)_3*9H_2O$, $Ga(NO_3)_3*7H_2O$, $Fe(NO_3)_3*9H_2O$, $Sr(NO_3)_2$, and glycerine. The water content were checked by TG measurements (Setaram Setsys evolution 1650) and found to be as given with the chemical formula. Appropriate amounts of the nitrates in order to obtain nominal compositions $Bi_2M_4O_9$ with M = Al, Ga, Fe as well as series $Bi_{2-2x}Sr_{2x}Al_4O_{9-x}$ and $Bi_{2-2x}Sr_{2x}Ga_4O_{9-x}$ (x = 0, 0.05, 0.1, 0.15) were dissolved in minimum amount of glycerin. The homogeneous solutions were slowly heated at 80°C to form a viscous gel. The gel was heated at 120°C in an oven for about 12 h. During this process the gel swells into a fluffy mass. The dried gel was calcined in an open Pt crucible at 850°C for 48 h.

The products were characterized by X-ray powder patterns recorded in a Stoe Stadi P diffractometer (transmission geometry, $CuK\alpha_1$ radiation by a focusing Ge (111) monocromator, linear PSD). Lattice parameter refinements and estimations of average crystal sizes were performed using the

Rietveld software Diffrac Plus TOPAS (Bruker AXS, Karlsruhe, Germany). For the calculation of the reflex profiles fundamental parameters were used on the basis of instrumental parameters calculated out of a silicon standard measurement. The zero point parameter was fitted and a polarization parameter was fixed.

Infrared absorption spectra were taken (FTIR spectrometer Bruker Vertex 80v) using PE and KBr pressed pellet technique in the spectral ranges $50 - 500$ $cm^{-1}$ and $370 - 4000$ $cm^{-1}$, respectively. Using the overlapping range the spectra were merged with respect to the KBr range with 1 mg sample dispersed in 200 mg KBr. Spectra are given in absorbance units ($-lg(I/I_0)$, $I_0$, I are transmitted intensities through "reference" and "reference + sample", respectively). $^{18}O/^{16}O$ exchange experiments were carried heating the as prepared samples at 800°C for 16 hours in $^{18}O_2$ atmospheric condition.

SEM/EDX (JEOL) were carried out using the as prepared and leached samples The leaching was done in 3 M $HNO_3$ for 1 h and then washed with destilled water and dried at 100°C.

## RESULTS AND DISCUSSION

$^{18}O/^{16}O$ exchange related diffusion coefficients in the basic systems $Bi_2M_4O_9$

It has been reported that the "Glycerin Method" reveal most easily single phase polycrystalline materials for the system $Bi_2(Al_xFe_{1-x})_4O_9$ [9, 13] compared to other methods using glycine combustion technique [4] or standard solid state method, as was used to prepare the series $Bi_2(Ga_xFe_{1-x})_4O_9$ [2]. Thus, the "Glycerine Method" was used to repeat the synthesis systematically for the two known systems and also to produce a new series of composition $Bi_2(Ga_xAl_{1-x})_4O_9$ with x = 0, 0.1, ..., 1.0 [9, 10]. It was observed that all the diffraction peaks of all products could perfectly be indexed to the orthorhombic structure (space group = Pbam). Small amounts of $Bi_2O_3$ and $Ga_2O_3$, if present, could be taken away by washing procedure. The initial values of the lattice parameters used for the refinements were taken from Müller-Buschbaum et al. [2] for $Bi_2Ga_4O_9$ and from Abrahams et al. [5] for $Bi_2Al_4O_9$. The refined lattice parameters for the end members $Bi_2M_4O_9$ (Tab. 1) well agreed with the reported values. Compared to earlier reports for the Fe/Al and Fe/Ga.

Table I List of lattice parameters of $Bi_2M_4O_9$ (M = Al, Ga, Fe) systems.

| System | Lattice parameters / Å | | |
|--------|------|------|------|
| | a | b | c |
| $Bi_2Al_4O_9$ | 7.71901(6) | 8.10511(7) | 5.68867(5) |
| $Bi_2Ga_4O_9$ | 7.92302(10) | 8.28952 (10) | 5.88880(7) |
| $Bi_2Fe_4O_9$ | 7.97663 (6) | 8.44235(5) | 6.00532(8) |

In Fig. 2, the IR absorption spectra of $^{18}O_2$ treated samples $Bi_2Al_4O_9$, $Bi_2Ga_4O_9$, and $Bi_2Fe_4O_9$ are compared with the spectra obtained before treatment. The spectra obtained for another $Bi_2Al_4O_9$ sample of significantly smaller crystal size is shown for comparison. It is observed that thermal treatment in $^{18}O_2$ reveal an additional peak (marked by arrow) which can be assigned to the vibration of the structure specific short $M-O_c$ bonding within the double tetrahedral unit $M_2O_7$. The peak positions for Ga-$^{18}O_c$, Al-$^{18}O_c$ and Fe-$^{18}O_c$ are 810, 881 and 771 $cm^{-1}$, respectively, i.e. about 41 $cm^{-1}$ lower compared to the appropriate peaks for Ga-$^{16}O_c$ (852 $cm^{-1}$), Al-$^{16}O_c$ (922 $cm^{-1}$) and Fe-$^{16}O_c$ (812 $cm^{-1}$). The isotope shift could be calculated using the measured value of $M-^{16}O_c$ peak according to

$$\upsilon_{M-^{18}O} = \upsilon_{M-^{16}O} \times (\mu_{M-^{16}O} / \mu_{M-^{18}O})^{1/2} \qquad (1)$$

($\upsilon_{M-O}$ and $\mu_{M-O}$ represent the peak position and reduced mass of M-O) revealing very good agreement with the experimental values.

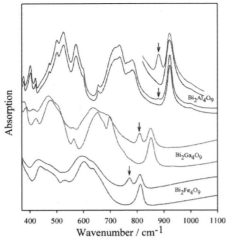

Fig. 2 FTIR absorption spectra of samples (as denoted) before and after $^{18}O_2$ treatment. The $M$-$^{18}O_c$ peaks are marked by arrow. Spectra of $2^{nd}$ $Bi_2Al_4O_9$ sample with smaller crystal size are also compared (top).

The $O_c$ is not the only oxygen site involved in the exchange reaction. It can be seen that all other peak positions are shifted to some extend and become broadened. These peaks cannot be further separated. However, the well separated $M$-$^{18}O_c$ related peak in the KBr-spectra enable an easy estimate of the $O_c$ related diffusion coefficient. For this the average crystal size calculated from X-ray data are given below in Tab. 2 together with the estimated relative peak intensities $M$-$^{18}O_c/M$-$^{16}O_c$. Assuming ball like crystals and the intensity ratio proportional to the volume ratio of unexchanged and exchanged part of the crystals the thickness of the exchanged outer shell z and the diffusion coefficients (z = sqrt(D*t), t = 57600 s) were calculated (Tab. 2). Although there are large uncertainties in our method which are in particular related to the values given for the average crystal size a right order of magnitude may be obtained. For $Bi_2Al_4O_9$ values from the two different batches reasonably agree. However values for $Bi_2Ga_4O_9$ for two different batches differ somewhat more. Here comparison to single crystal data could be given [14], where a diffusion coefficient of D = $2*10^{-14}$ $cm^2$/s was obtained as in SIMS depth profile measurement on a (010) plate from the single crystal isotope exchange experiments at the same temperature. This might be due to a systematic failure in our method considering only the diffusion effect observed for the $O_c$ oxygen. This could also imply that this oxygen does not contribute an extraordinary high value. The diffusion coefficient obtained for $Bi_2Fe_4O_9$ ($5*10^{-17}$ $cm^2$/s) is in the same magnitude compared to the bismuth gallate.

$^{18}O/^{16}O$ exchange related diffusion coefficients in the sytems $Bi_{2-2x}Sr_{2x}M_4O_{9-x}$, M = Ga, Al

The "Glycerin Method" was used also to prepare $Bi_{2-2x}Sr_{2x}M_4O_{9-x}$ with M = Ga, Al and x = 0, 0.05, 0.1 and 0.15. XRD data of the products showed reasonably pure single mullite type phase with some negligible low intensity additional peaks which could be related to impurity phase. The most interesting question was how the diffusion coefficient develops with nominal increase of Sr. Therefore, the IR absorption spectra are considered first. For $Bi_{2-2x}Sr_{2x}Al_4O_{9-x}$ (Fig. 3) there are only small changes in the spectra with increasing x, which are related mostly to a decreasing intensity of all peaks. An additional broad peak (called Al* in the following) appears at about 844 $cm^{-1}$, which increases systematically with increasing x as the intensity of the Al-$^{16}O_c$ related peak at 922 $cm^{-1}$ decreases. This could imply a significant loss of Al-$^{16}$O$_c$ related oscillator strength which becomes transferred into the Al* peak. Sintering at 800°C for 4 days even led to a slight sharpening of the Al* peak. But it was also observed that heating for longer time (14 days) led to a disappearance of Al* [12]. Heating of the as prepared

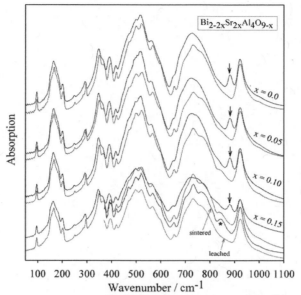

Fig. 3 FTIR absorption spectra of samples Bi$_{2-2x}$Sr$_{2x}$Al$_4$O$_{9-x}$ before and after $^{18}$O$_2$ treatment. The Al-$^{18}$O$_c$ peaks are marked by arrow. The Al* peaks are marked by * (for detail see text).

samples in$^{18}$O$_2$ atmosphere for 16 h at 800°C reveal a strong well separated Al-$^{18}$O$_c$ absorption peak and significant shift in all other peaks. The smallest shifts are observed for the low wavenumber peaks which are related to Bi-O vibrations due to the high contribution of Bi to the force constants. As described above the peak intensity ratio Al-$^{18}$O$_c$/Al-$^{16}$O$_c$ was evaluated revealing, however, almost unchanged diffusion coefficients with respect to the x = 0.0 composition (Tab. 2). Considering the Bi$_{2-2x}$Sr$_{2x}$Ga$_4$O$_{9-x}$ system (Fig. 4) a similar behaviour is observed. In particular a clearly visible additional peak (called Ga* in the following) is observed at about 789 cm$^{-1}$ which systematically increases with increasing x whereas the Ga-$^{16}$O$_c$ peak decreases in intensity. Heating of the as prepared sample in $^{18}$O$_2$ atmosphere for 16 h at 800°C reveal a strong well separated Ga-$^{18}$O$_c$ absorption peak and significant shift in all other peaks. As described above the peak intensity ratio Ga-$^{18}$O$_c$/Ga-$^{16}$O$_c$ was evaluated revealing almost unchanged diffusion coefficients with respect to the x = 0.0 composition (Tab. 2). It may be noted that the Ga* peak shows a systematic shift towards lower wavenumber related to the $^{18}$O$_2$ treatment. The peak becomes broader and loose some intensity.

Table II Diffusion coefficients (D) at 800°C, calculated from the FTIR absorption spectra of $^{18}$O$_2$ treated samples. R= IR peak intensity ratio, z = outer layer thickness.

| Sample number | Compositions (nominal) | Average crystal size, (d/2)/nm | R = M-$^{18}$O$_c$ /M-$^{16}$O$_c$ | z/nm | D/cm$^2$/s |
|---|---|---|---|---|---|
| 1 | Bi$_2$Al$_4$O$_9$ | 196 | 0.05 | 3.2 | 1.7×10$^{-18}$ |
| 2 | Bi$_2$Ga$_4$O$_9$ | 197 | 0.62 | 29.2 | 1.6×10$^{-16}$ |
| 3 | Bi$_2$Fe$_4$O$_9$ | 113 | 0.67 | 17.8 | 5.0×10$^{-17}$ |
| 4 | Bi$_2$Al$_4$O$_9$ | 35 | 0.39 | 3.6 | 2.3×10$^{-18}$ |
| 5 | Bi$_{1.9}$Sr$_{0.1}$Al$_4$O$_{8.95}$ | 30 | 0.47 | 3.6 | 2.3×10$^{-18}$ |
| 6 | Bi$_{1.8}$Sr$_{0.2}$Al$_4$O$_{8.9}$ | 29 | 0.44 | 3.3 | 1.9×10$^{-18}$ |
| 7 | Bi$_{1.7}$Sr$_{0.3}$Al$_4$O$_{8.85}$ | 27 | 0.43 | 3.0 | 1.6×10$^{-18}$ |
| 8 | Bi$_2$Ga$_4$O$_9$ | 137 | 0.48 | 16.8 | 4.8×10$^{-17}$ |
| 9 | Bi$_{1.9}$Sr$_{0.1}$Ga$_4$O$_{8.95}$ | 42 | 0.62 | 6.2 | 6.7×10$^{-18}$ |
| 10 | Bi$_{1.8}$Sr$_{0.2}$Ga$_4$O$_{8.9}$ | 46 | 0.52 | 6.0 | 6.2×10$^{-18}$ |
| 11 | Bi$_{1.7}$Sr$_{0.3}$Ga$_4$O$_{8.85}$ | 38 | 0.56 | 5.2 | 4.7×10$^{-18}$ |

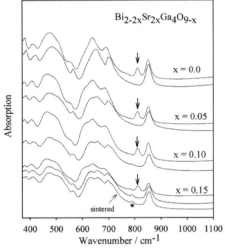

Fig. 4 FTIR absorption spectra of Bi$_{2-2x}$Sr$_{2x}$Ga$_4$O$_{9-x}$ samples, before and after $^{18}$O$_2$ treatment. (Symbols as described in Fig. 3).

For all as prepared samples acid leaching experiments were carried out. The IR absorption spectra then showed in all cases agreement with the corresponding x = 0 samples including the complete absence of the Al* and Ga* denoted peaks. The assignment as Al*, Ga* should indicate the suggested relation to the mullite specific T* site due to the formation of vacant O$_c$ atoms which also lead to special peak in the IR absorption spectra of mullite [15]. However, presently there is no clear evidence that Al* and Ga* can be related to the mullite type structure. This will be further investigated in the following using XRD and SEM/EDX methods.

XRD and SEM/EDX analysis of acid leached and unleached samples

The XRD pattern of the as prepared samples Bi$_{2-2x}$Sr$_{2x}$Al$_4$O$_9$ with x = 0.0, 0.05, 0.10 and 0.15 together with the acid leached sample x = 0.15 are shown in Fig. 5. The leaching result shows the disappearance of all extra peaks which could not be refined within space group Pbam.

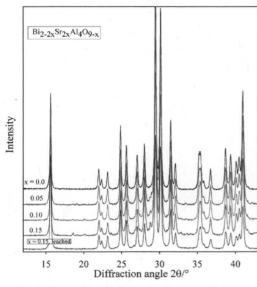

Fig. 5 XRD patterns of samples with nominal compositions Bi$_{2-2x}$Sr$_{2x}$Al$_4$O$_9$.x. Few weak diffraction lines appear with increasing nominal x which, however, disappear after leaching in acid (bottom).

It also indicates the inhomogenous nature of the sample with respect to composition. It has been checked by EDX analysis that the detectable Sr content nearly completely disappeared by leaching, whereas the Sr content systematically increases with increasing x for the as

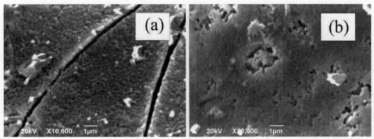

Fig. 6 SEM images of sample with nominal composition Bi$_{1.7}$Sr$_{0.3}$Al$_4$O$_{8.85}$ (a) as prepared sample and (b) sample after leached in acid.

prepared samples. Therefore the main Sr containing part of the sample could be seen as a surface layer to the $Bi_2Al_4O_9$ nanocrystals and can be incorporated only to a very limited content (less than 5%) in the bulk. SEM pictures (Fig. 6) comparing the acid treated and as prepared sample showed with the presently available resolution mostly no resolvable grains and some larger pieces for the acid treated and as prepared sample, respectively. Analogous results were obtained for the $Bi_{2-2x}Sr_{2x}Ga_4O_{9-x}$ system.

SUMMARY AND CONCLUSION

Series of composition of $Bi_{2-2x}Sr_{2x}M_4O_{9-x}$ with M = Al and Ga were prepared using the Glycerin Method. Lattice parameter refinements indicated no significant variation in values related on x with respect to the x = 0 compositions. Therefore, from this no conclusion about any doping effect could be drawn. Further Rietveld structure refinements could imply that Sr could substitute Bi. However, the main Sr content belongs to impurity phases. Evidence for this could be shown by acid leaching experiments, revealing both in the IR absorption spectra as well as the XRD pattern agreement with the spectra of the x = 0 composition. Additionally EDX showed the nearly complete absence of Sr after leaching whereas the as prepared samples showed a systematic increase in Sr-content with increasing x. Moreover calculations of the average crystal sizes showed no significant change for the acid leached samples compared to the as prepared ones. SEM pictures could give the impression of largely intergrown crystals in the as prepared samples, whereas for the leached samples no contours could be separated. Therefore it may presently not be ruled out that the nanocrystals are intergrown containing Sr rich interfaces or surfaces, which shows a close relationship to mullite structure. A hint for this could be related to the observation of T* peaks in the infrared absorption. Following this line of arguments, the Sr containing surfaces are highly permeable for oxygen. However, the bulk crystal shows low oxygen diffusivity as obtained by the $^{18}O/^{16}O$ exchange via infrared absorption.

ACKNOWLEDGEMENTS

The authors thank the 'Deutsche Forschungsgemeinschaft' (DFG) for financial support (GE 1981/2-1; PAK 279).

REFERENCES

[1] R.X. Fischer, H. Schneider, "The mullite-type family of crystal structures", in: H. Schneider, S. Komarneni (Eds.), *Mullite*, Wiley-VCH, Weinheim 2005, pp.1-140.
[2] H. Müller-Buschbaum, D.C. de Beaulieu, "About occupation of octahedral positions and tetrahedral positions of $Bi_2Ga_2Fe_2O_9$," *Z. Naturforsch. B Chem. Sci.* **33** (1978) 669-670.
[3] D.M. Giaquinta, G.C. Papaethymiou, W.M. Davis, H.-C. Zur Loye, "Synthesis, structure, and magnetic-properties of the layered bismuth transition-metal oxide solid-solution $Bi_2Fe_{4-x}Ga_xO_9$," *J. Sol. State Chem.* **99** (1992) 120-133.
[4] S.W. Zha, J. G. Cheng, Y. Liu, X. G. Liu, G. Y. Meng, "Electrical properties of pure and Sr-doped $Bi_2Al_4O_9$ ceramics," *Solid State Ionics* **156** (2003) 197-200.
[5] I. Abrahams, A.J. Bush, G.E. Hawkes, T. Nunes, "Structure and oxide ion conductivity mechanism in $Bi_2Al_4O_9$ by combined X-ray and high-resolution neutron powder diffraction and Al-27 solid state NMR," *J. Solid State Chem.* **147** (1999) 631-636.
[6] J. Schreuer, M. Burianek, M. Muhlberg, B. Winkler, D. J. Wilson, H. Schneider, "Crystal growth and elastic properties of orthorhombic $Bi_2Ga_4O_9$," *J. Phys. Condens. Matter* **18** (2006) 10977-10988.
[7] L. López-de-la-Torre, A. Friedrich, E.A. Juarez-Arellano, B. Winkler, D.J. Wilson, L. Bayarjargal, M. Hanfland, M. Burianek, M. Mühlberg, H. Schneider, "High-pressure behavior of the ternary bismuth oxides $Bi_2Al_4O_9$, $Bi_2Ga_4O_9$ and $Bi_2Mn_4O_{10}$," *J. Solid State Chem.* **182** (2009) 767-777.

[8] S. Larose, S.A. Akbar, "Synthesis and electrical properties of dense $Bi_2Al_4O_9$," *J. Solid State Electrochem.* **10** (2006) 488-498.

[9] T. Debnath, C.H Rüscher , P. Fielitz, S. Ohmann, G. Borchardt, "Synthesis and characterisation in the new system $Bi_2(Al/Ga)_4O_9$ and $^{18}O/^{16}O$ exchange experiments," 8th Pacific Rim Conference on Ceramic and Glass Technology, Vancouver (Canada) May 31 – June 5, 2009.

[10] T. Debnath, C.H. Rüscher, "Crystal structure of bismuth aluminium gallium oxide, $Bi_2(Ga_xAl_{1-x})_4O_9$, x = 0.4, 0.6, 0.8," *Z. Kristallogr. NCS.* DOI 10.1524/ncrs.2010.0001.

[11] C.H. Rüscher, T. Debnath, P. Fielitz, S. Ohmann, G. Borchardt, "$^{18}O/^{16}O$ exchange studies in the new series of compositions $Bi_2(Ga_xAl_{1-x})_4O_9$: Diffusion coefficients estimated from infrared absorption and XRD powder data," *Diffusion Fundamentals* **11** (2009) p9.1-p9.2.

[12] C.H. Rüscher, Th.M. Gesing, J.-Chr. Buhl, T. Debnath, "Synthesis and characterisation in the new system $Bi_{2-2x}Sr_{2x}Al_4O_{9-x}$ and $^{18}O/^{16}O$ exchange experiments," *Z. Kristallogr. Suppl.* **29** (2009) 88.

[13] Th.M. Gesing, C.H. Rüscher, J.-Chr. Buhl, "High-temperature X-ray diffraction study of mullite type $Bi_2M_4O_9$ phases (M = Al, Ga)," *Z. Kristallogr. Suppl.* **29** (2009) 84.

[14] P. Fielitz et al, will be published later.

[15] C.H. Rüscher, "Thermic transformation of sillimanite single crystals to 3 : 2 mullite plus melt: investigations by polarized IR-reflection micro spectroscopy," *J. Eur. Cer. Soc.* **21** (2001) 2463-2469.

AQUEOUS PROCESSING FOR SELF STANDING YSZ FILMS FOR SOFC STUDIES

Srinivasan Ramanathan
Materials Processing Division
BARC, Mumbai, Maharastra, India

ABSTRACT

Formation of flat sintered YSZ films (0.05 to 0.5mm thick and diameter~50mm) by aqueous tape casting using nano-crystalline powder prepared by reverse strike co-precipitation method was studied. TG-DTA and XRD studies revealed the optimum calcination temperature to be $900^0$C. The dry ground powder was found to be agglomerated and was wet ground under an optimized dispersion condition to form into a slurry with agglomerate size of $D_{50} \sim 0.7\mu m$. The slurries for tape casting were formulated using a polyvinyl alcohol-glycerol-water based binder solution and the composition of the ingredients was optimized through a detailed rheological study. Slurries were formulated with desired amount of pseudoplasticity that could exhibit controlled flow to form flexible films with desirable thickness. The binder burn out schedule and sintering conditions were optimized to form flat and dense electrolyte films (>98%T.D.) The sintered film exhibited fine grained microstructure.

INTRODUCTION

Solid oxide fuel cells are considered as potential electrical energy sources due to their higher efficiency and release of environmental friendly gaseous end products. Among the various designs of fuel cells, planar type is claimed to have the advantages of possessing higher power density per unit volume and low production costs.[1] Dense self standing thin films (100 to 500µm) of yttria stabilized zirconia (YSZ) finds application as the electrolyte in planar test cells used for carrying out cell performance studies such as current-voltage (i-v) characteristics as a function of composition, micro-structure, thickness etc. of itself and electrodes (anode & cathode) in contact with it.

Aqueous based slurry processing for formation of thin YSZ film is becoming important as the processes are cheap and environment friendly.[2-4] The various stages in ceramic processing of these self standing films are preparation of precursor powder and powder treatment (calcination and grinding), evaluation of their sintering behavior, formulation of well dispersed slurry with binder, casting, controlled binder burn out and sintering.

Among the various methods reported for the synthesis of YSZ powders such as decomposition of mixed nitrates, solution combustion and co- precipitation the most widely studied and used process is coprecipitation etc.[5] In co-precipitation different metal ions are precipitated together resulting in compositional homogeneity or mixing in molecular level.

The precursor formed by co-precipitation require to be evaluated for its thermal decomposition and phase evolution behavior to fix the calcination temperature. The powder formed by calcination need to be further treated to obtain optimum characteristics for sintering bodies with desired microstructure. The powder formed are generally agglomerated which need to be eliminated by wet grinding to improve the green microstructure and sinter-ability to form fine grained bodies. The conditions for wet grinding and tape casting are optimized through zeta-potential and rheology studies. Defect free flat films are formed by controlled binder burn out followed by sintering. Detailed investigation starting from preparation of powder to formation of sintered films by aqueous tape casting has not been reported much for YSZ systems. The processing for preparation of sinter-active YSZ powders using co-precipitation method has been studied in our laboratory.[5] Formation of self standing films using this powder by aqueous tape casting has been recently studied by us and the results are reported in this paper.

EXPERIMENTAL

The precursor (corresponding to a 30g batch of YSZ powder) was prepared by the reverse strike co-precipitation of hydroxides by drop-wise addition of a mixture of the aqueous solution containing required amount of zirconium oxy-chloride and yttrium nitrate (0.1M metal concentration) in to a bath containing excess ammonia (0.1M) with vigorous stirring. Excess amount (i.e., double the amount required) of ammonia was used to keep pH around 10 through out precipitation. The precipitate was washed for five times with distilled water. While washing, the precipitate was stirred with same amount of water at a pH of 8 for half an hour and was allowed to settle down. This enabled equilibration of the entrapped chloride with the bulk and pH of 8 prevented precipitation. After washing for five times, the filtrate was found to be free of chloride as shown by the silver nitrate test. Final washing of the precipitate was done with alcohol by allowing the precipitate to remain in the alcohol overnight for effective exchange of water and alcohol to medium and gel respectively. The precipitate after filtration was dried in air and then in air oven at $80^0$C. The powder thus formed was homogenized by dry grinding in a planetary mill.

The as formed powder was subjected to TG-DTA and XRD characterization to fix the calcinations temperature. The powder calcined at $900^0$C was dry and wet ground in a planetary mill (using 30g of the powder in a pot of 80cc volume and 15 number of alumina balls of 10 mm diameter at a mill speed of 200 r.p.m). The ground powders were characterized for agglomerate size distribution by laser light scattering technique.. Dispersion conditions for slurry formulation for effective wet grinding and quality tape casting were optimized using zeta potential and viscosity measurements. Zeta-potential variation with pH for aqueous suspensions of powders was studied using laser Doppler velocimetry. Rheological characteristics of the slurries were evaluated using a cone and plate type viscometer.

Tape cast slurries were prepared by wet mixing the submicron sized powder at a pH of 3 (the pH of maximum zeta potential) followed by addition of polyvinyl binder solution (polyvinyl alcohol: glycerol: water = 10:5:85) and further mixed in the ball mill. Tape casting of sheet specimens was carried out with a blade gap of 300 to 800 μm using a laboratory assembled set up. Cast tapes were dried overnight and cut to desired sizes (up to ~5cm) . The tapes were subjected to controlled binder burn out (rate of heating of $0.5^0$C /minute up to $600^0$C in a Kanthal furnace in air) followed by final sintering to $1500^0$C (rate of heating of $5^0$C / minute in a Super-Kanthal furnace). The films were sandwiched in between flat zirconia fiber boards to improve their flatness after sintering. Sintered specimens were characterized for density by the Archimedes method and microstructure by SEM.

RESULTS AND DISCUSSION

a). Precursor formation

As co-precipitation involves simultaneous precipitation of hydroxides of different metals possessing different solubilities, concentration of the precipitant (i.e., pH of the medium) should be more than that required to precipitate both zirconium and yttrium simultaneously to improve compositional homogeneity. Using excess amount of ammonia and reverse strike method ensured co-precipitation of both $Zr^{4+}$ and $Y^{3+}$ ions. It was found that the effectiveness of washing improved with prolonged stirring and stay in wash medium as shown by the silver nitrate test for the filtrate. As the equilibration with the wash medium takes long time the precipitate was digested in the alcohol overnight. After oven drying, the voluminous gel became a compact and easily friable powder cake which was planetary milled to form in to a homogeneous mass.

b). Phase evolution and wet grinding

The XRD patterns of the as formed powders and that calcined at 900°C are shown in Fig.1. The oven dried precipitate was amorphous while that heated to 900°C is crystalline cubic phase (pc-pdf no.YSZ: 30-1468). Molecular level mixing in solution based techniques is attributed to result in nanocrystalline phase pure material formation. The TG-DTA pattern for the precursor is shown in Fig.2. The YSZ gel exhibited about 25wt% accompanied by a broad endotherm in the temperature range of 25 to 400°C. This is attributed to the loss of water, both absorbed and entrapped in the capillaries of the gel.[6] There was a sharp exotherm around 450°C which is attributed to the crystallization of cubic YSZ phase. Even though the exotherm was complete below 500°C, there was a slight loss of weight till 900°C while the crystallite size did not increase much (5 to 10nm only). Hence the powder was calcimined at this temperature.

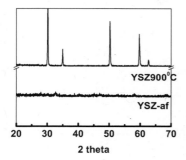

Fig.1 XRD patterns for precursor & that heated to 900°C

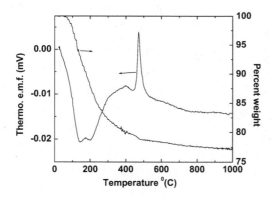

Fig. 2. TG-DTA patterns for precursor

The calcined powder, upon dry grinding, formed a cake along the walls of the pot in half an hour and hence further milling is expected not to contribute to size reduction. The particle size distribution of the dry ground powder as shown in Fig.3 exhibits a broader size distribution with a $D_{50}$ of 6$\mu$m with substantial fraction remaining in the size above 10$\mu$m. As presence coarse agglomerates hinder sintering and create microstructural heterogeneity the powder was further wet ground. A detailed investigation on the dispersion characteristics and its role on the grinding behavior of this powder has been carried out by us and reported elsewhere.[5] The wet grinding of the powder was carried out using a pH of 3 (point of maximum zeta potential) and solid content of 20 vol% (near Newtonian flow) for 10hrs. The particle size distribution of the wet ground slurry, exhibited a mean size ($D_{50}$ ) of 0.7$\mu$m and absence of coarse agglomerates. Such a powder is expected to yield well sintered bodies with homogeneous microstructure. Wet ground sub-micron sized ($D_{50}$ ~ 0.5 to 0.7$\mu$m) powders of YSZ calcined at a temperature of 900$^0$C resulted in formation highly dense (95% T.D.) electrolyte bodies upon sintering at 1400$^0$C. [05] The 900$^0$C calcined powder, possess optimum amount of sinter-activity there by retaining the desired amount of thermodynamic driving force available for sintering while the finer particle size decreased the distance over which mass transport has to take place thereby increasing the kinetics of sintering.

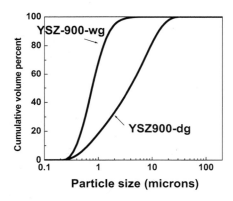

Fig.3. Particle size distribution of the powders used

c). Zeta potential studies

For formation of quality aqueous slurry for tape casting, the precondition is formation of stable powder dispersion in the water. One method of achieving dispersion of particles in aqueous medium is through the electrostatic repulsion due to presence of charge on their surface, which is measured through the zeta potential. Houivet et al. have presented a detailed description of the origin of zeta potential in oxide suspensions.[7-9] A value greater than 20mV is an indication of the deflocculated state and higher the value better is the dispersion stability. There is a wide variation in the reported zeta potential values for each powder system depending upon the method of its preparation (due to presence of trace amount of impure ions on the surface of the powders) and calcination condition (due to variation in the reactivity of the surface with water and the distilling away of the impure ions). Hence it is essential to find out actual values for these powders prepared in our laboratory and hence has been carried out for suspensions of all these powders as a function of pH. The variation in zeta potential of the aqueous suspensions with pH is shown in Fig.4. and it is noted that it exhibits a maximum in both acidic (pH3) and alkaline (pH 10) ranges. Also, the numerical value of maximum is higher in acidic pH and hence the slurries for grinding and tape casting were formulated at a pH of 3.

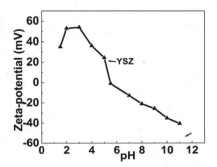

Fig.4. Zeta potential variation with pH for the aqueous suspension of YSZ

d).Rheological characterization

Even at the pH of maximum zeta potential, flocculates are reported to exist in concentrated slurries and it increases with increasing solid loading and binder addition. It is indicated by their rheological behavior (pseudo-plasticity or shear thinning behavior).[2,10] It is undesired for effective wet grinding cum mixing of the ingredients (binder, plasicizer etc) and desired from the point of view of casting tapes with required thickness. Also, it is not possible to completely eliminate them and always optimum conditions are arrived at from the point of view of processing requirements (eg. Newtonian flow for wet grinding & pseudoplastic flow for tape casting). The viscosity variation with shear rate and viscoelastic behavior of slurries of varying compositions has been studied.

A pH of 3 was chosen for slurry formulation due to presence of sufficient amount of charge on the particles (45mV at this pH. The viscosity variation with shear rate plot (Fig.5) for the slurry containing 30vol% solid content exhibited pseudo-plasticity i.e., decrease in viscosity with increase in shear rate. Pseudo-plasticity generally indicates presence of flocculates which entrap the water and make it immobile leading to higher viscosity. As they are broken upon shearing, the water becomes available for reducing the viscosity. Thus with increasing shear rate, the viscosity decreases.

A detailed study of the tapes formed with YSZ slurries containing varying amounts of binder-plasticizer solution was carried out and it was found that a slurry containing about 10 weight percent of binder with respect to powder content was required for formation of flexible tapes that could be peeled off. Hence rheological behavior of slurries containing above-mentioned amount of powder-binder combination with varying amount of water was studied and the results are given in Fig 5. All the slurries exhibited shear thinning (i.e., decrease of viscosity with increasing shear rate) flow behavior. The viscosity of the slurry with out binder solution was found to be the lowest even though the solid loading is the highest (30vol%). Slurries with binder, in spite of lowering of solid content, exhibited higher viscosity which is attributed to binder-powder-water interaction. Actually, the solid content decreased with addition of binder solution. The viscosity variation with shear rate plots for these slurries also exhibited pseudo-plasticity. It is interesting to note with slight decrease in solid content (~ 2wt%), the viscosity decreased drastically. Slurries-C and B (18 & 20vol.% solid content) exhibited time independent flow behavior (pseudoplasticity) while slurry-A (22vol.% solid content) exhibited time dependent variation in viscosity during increase and decrease of shear rate (thixotropy). Thixotropy is not desirable for formation of quality casting as flow characteristics vary during the time of casting. In addition, at this solid content, it is difficult to mix and de-air the slurry. Uncontrolled flow of slurry occurs with slip of lower solid contents (slurries with solid content lesser than that corresponding to 'C'). The slurry-B was found to be the optimum one for formation of quality casts. The viscosity was high enough to avoid sedimentation of particles during drying of the wet tapes and still low enough for easy removal of air bubbles and exhibit controlled flow to form castings with desired thicknesses. The results of the visco-elastic characterization studies on these slurries is shown in Fig.6. In amplitude sweep mode, the slurries were found to be predominantly viscous as the viscous modulus component was always higher and remained constant with varying shear rate. But the elastic modulus decreased with increasing shear rate above a critical value indicative of the breaking away of the internal structure. With increasing solid content the critical shear increased indicating higher degree of interactions in concentrated slurries. The phase angle value also was less than 90 at lower shear stress due to presence of elastic component and it increased to 90 degree with higher shearing increasing characteristic of a pure viscous liquid. . All these results indicated the increased degree of association that arises due to binder particle medium interactions.

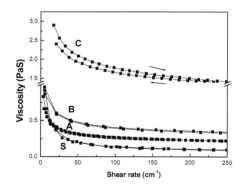

Fig.5. Variation of viscosity with shear rate for slurries
S-YSZ in water (30vol.% at pH 3), A, B, C – slurries (YSZ-Binder-water)- with 18, 20, 22vol% solid in water

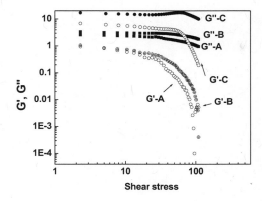

Fig.6 . Variation of elastic (G') and viscous modulii (G'') with increasing amplitude of shear stress for tape cast slurries

e). Tape casting, binder burn out, sintering and characterization of the tapes

The tapes were cast with wet slurry thickness varying from 300 to 800 µm which yielded tapes with dried thickness in the range of 100 to 500 µm. As the binder burned in the temperature range of 250 to $600^0$C, tapes were heated at a slower rate of heating (0.5$^0$C per minute) up to $600^0$C to avoid formation of cracks due to evolution of gases. Further heating was done at a faster rate (5$^0$C per minute until 1500$^0$C and soaked for 2 hours). The sintered specimens of electrolyte exhibited densities value greater than 98%T.D and the typical microstructures of the sintered specimen exhibited fine grained structure (Fig.8).

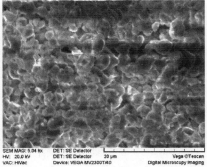

Fig.8.Typical microstructure of the as fired surface of the sintered tape

Fig.9. Flat sintered film (thickness 0.2 mm)

CONCLUSIONS

The precursor for YSZ was formed by aqueous co-precipitation technique and conditions for formation of phase pure powder was optimized through XRD and TG- DTA studies. The deagglomeration of the powder to submicron size (0.7micron) was carried out by wet grinding. Dispersion conditions for formation of aqueous slurries were optimized through zeta-potential and viscosity measurements. Shear rate dependent viscosity variation and viscoelastic characterization studies revealed the presence of interaction between particle-binder-water with in the slurries Conditions were optimized to form flat self standing dense films by sintering. The sintered tapes of YSZ exhibited fine grain microstructure.

ACKNOWLEDGEMENTS
The author thanks Dr N.Krishnamurthy, Head High Temperature Processing,  Dr I.G. Sharma, Head Materials Processing and Dr.A.K. Suri, Director Materials Group for their interest in the work.

REFERENCES
1. Masashi Mori and Yoshiko Hiei,  J. Am. Ceram. Soc. , 84,  11, 2573-78 (2001).
2. Bernd Bitterlich, Christiane, Lutz, Andreas Roosen, Ceramics International, 28, 675-683 (2002)
3. P. Snijkers, A de Wilde, S. Mullens, J. Luyten J. European Ceram. Soc, 24, 6, 1107-1110
4. D. Hotza, P. Greil, , Mater. Sci. Eng. A202, 206-217 (1995)
5. S. Ramanathan, K.P.Krishnakumar, P.K. De, S. Banerjee,Bull. Mater.Sci. 28, 2, 109-114 (2005)
6.S. Ramanathan, R.V. Muraleedharan, S.K. Roy, P.K.K. Nair J. Am. Ceram. Soc.78, 429 (1995)
7.David Houivet, Jaffar El Fallah and Jean Marie Haussonne, J. Am. Ceram. Soc., 2002, 85, 2, 321 – 28
8.Fries Robert, Rand Brian, Materials Science and technology, processing of ceramics, Part I, Eds. R. W. Cahn, Weinheim:VCH Publishers Inc., Vol.17A, p.153 (1996)
9.Hellobrand Hans,  Materials Science and technology, processing of ceramics, Part I, Eds. R. W. Cahn, Weinheim:VCH Publishers Inc., Vol.17A, p.189 (1996)
10.Amit Mukherjee, B. Maiti, A. DAS Sharma, R.N. Basu, H.S. Maiti, Ceramics International, 27, 731-739 (2001)

# SYNTHESIS AND SINTERING OF YTTRIUM-DOPED BARIUM ZIRCONATE

Ashok K. Sahu, Abhijit Ghosh, Soumyajit Koley and Ashok K. Suri
Materials Group, Bhabha Atomic Research Centre, Mumbai – 400085 (India)

ABSTRACT

Ba(Zr$_{0.85}$Y$_{0.15}$)O$_{3-\delta}$ (BZY) has been synthesized by co-precipitation method. Powder x-ray diffraction patterns confirmed phase pure material formation at 1100°C. The average particle size of the 1100°C calcined powder was found to be ~1μm. The nature of the powder was found to be softly agglomerated. Shrinkage behaviour of powder compacts has been studied.

Effect of addition of different amounts of ZnO on the shrinkage behaviour and sintered density of BZY has also been studied. It was observed that addition of ZnO enhanced sintering, sintered density of more than 95 % TD was achieved by sintering at 1400°C. Study of sintering kinetics of powder compacts with 1 wt % ZnO and without ZnO has been carried out. Initial results show that different sintering mechanisms operate in these two materials. This is also supported by microstructural studies.

## I. INTRODUCTION

Barium zirconate (BZ) with a perovskite structure is of interest for variety of applications. It is a promising container material for the manufacture of high temperature superconductors as it is chemically compatible with the liquid phases produced at high temperature during the process[1]. Yttrium/scandium doped barium zirconate have attracted attention for their potential applications as electrolytes in fuel cells[2]. For such applications, highly dense impervious ceramic parts are required. However, the refractory nature of doped BZ makes it difficult to process for such applications. For sintering barium zirconate, extreme conditions, such as high temperature (1700°C), long sintering times (24 hrs), and nanopowders are required to prepare fully dense products[3]. Hence it was considered worthwhile to explore the possibility of making BZ with suitable dopant/additive to lower down their sintering temperature. An attempt has also been made to study the sintering kinetics of powder compacts with and without the addition of ZnO.

## II. EXPERIMENTAL

### (1) SYNTHESIS OF POWDER

Yttrium doped barium zirconate, Ba(Zr$_{0.85}$Y$_{0.15}$)O$_{3-\delta}$ (BZY) has been synthesized by reverse strike coprecipitation method. Solutions of BaCO$_3$ in acetic acid, Y$_2$O$_3$ in nitric acid and ZrOCl$_2$.8H$_2$O in water were prepared separately. The strength of each individual solution was approximately 0.5M. A mixed solution was prepared containing appropriate amounts of cations. The mixed solution was added drop wise, with continuous stirring, to a precipitating bath containing ammonium carbonate and ammonium hydroxide. A 100 % excess ammonium carbonate compared to the stoichiometry for barium carbonate formation was taken to ensure complete precipitation of barium. The pH and temperature of the bath were maintained at around 9 and 50°C respectively. The precipitate was easily filterable and washed several times with ammoniacal water till it was free of chloride ions. Finally it was washed with alcohol. The wet cake was dried overnight in an oven at 80°C.

### (2) POWDER CHARACTERIZATION

The phase identification of the precursor was performed by x-ray diffraction (XRD). The precursor powder was calcined at various temperatures in the temperature range of 700 – 1100°C. Powder diffraction patterns of all the calcined powders were recorded. The powder calcined at 1100°C was ground for 3 hrs in a planetary mill using zirconia pot and balls. Acetone was used as the medium for grinding. Particle size distribution of the ground powder was recorded in a CILAS 1064 particle size analyzer.

To study the effect of ZnO on sintering of BZY, 0.5, 1, 2, and 4 wt % ZnO were added to the calcined powder before grinding.

## (3) POWDER COMPACTION AND SINTERING

The 1100°C calniced and ground powder was uniaxially pressed at room temperature using 9.5 mm diameter die-punch assembly at different compaction pressures upto 300 MPa. The uniaxially pressed pellets were then subjected to cold isostatic pressing at pressures upto 300 MPa using an EPSI isostatic press to achieve uniform density distribution throughout the pellet. The densities of the compacted green pellets were measured from the mass and the geometry of the sample. To find out the nature of agglomeration in the powder, which is determined by the agglomeration strength of the powder, green density was plotted against compaction pressure. The pellets pressed at 300 MPa were sintered at 1300 and 1400°C. BZY with various wt % ZnO were also sintered. In the case of sintered pellets, the Archimedes' principle was used for measuring the density.

Shrinkage behaviour of both the powder compacts, with and without ZnO isostaticaly pressed at 300 MPa was studied with the help of a dilatometer, model TD 5000S, MAC Science, Japan. Shrinkage behaviour was recorded in the temperature range 100 – 1500°C.

For sintering kinetics study, constant rate of heating experiment was carried out with BZY and BZY with 1 wt % ZnO. Green compacts, pressed isostatically at 300 MPa, were used for recording linear shrinkage during heating with a constant rate (300°C/h) upto 1500°C in air. The approach for the data analysis was based on Johnson model[4] treating grain boundary diffusion (GBD) and volume diffusion (VD) separately. This assumption is valid only at the initial stage of sintering (5% shrinkage). For CRH experiment (T=ct where "c" is the heating rate), the general form of modified Johnson equation can be written as

$$y^{(\frac{1}{n}-1)}T\frac{dy}{dT} = A\exp\left(-\frac{Q}{RT}\right) \qquad (1)$$

where, y = relative shrinkage (=$\Delta L/L_o$, where $\Delta L$ is shrinkage = $L_o$-$L_t$, $L_o$=initial length and $L_t$= length at time t ), dy/dt is the rate of shrinkage, A is a constant, Q is the activation energy, T is the temperature (K) and R is the gas constant. The value of n determines the sintering mechanism.

After integration and differentiation of Eq.1, Young and Cutler[5] made the following equation (Eq.2):

$$T\frac{dy}{dT} \approx A\exp\left(-\frac{nQ}{RT}\right) \qquad (2)$$

Based on this equation, they concluded that a plot of $\ln\left[T\frac{dy}{dT}\right]$ vs. $\left(\frac{1}{T}\right)$ would give a straight line and slope of that straight line would be $\left(-\frac{nQ}{R}\right)$. Woolfrey and Bannister[6] proposed to plot $\left[T^2\frac{dy}{dT}\right]$ vs. 'y' to obtain a slope of $\left(\frac{nQ}{R}\right)$. The sintering mechanism (n) was determined using these two models, by substituting the Q value obtained from the Dorn method[7]. In this method, the instantaneous effect on shrinkage rate of a small step change in temperature is determined. If $\left[\frac{dy}{dT}\right]_{T_1}$ is the shrinkage rate at a temperature $T_1$ (K) just before the temperature change and

$\left[\dfrac{dy}{dT}\right]_{T_2}$ is the rate of shrinkage at temperature $T_2$ just after the change, the activation energy of the process responsible for sintering is given by (Eq.3):

$$Q = \frac{RT_1T_2}{T_1 - T_2} \ln\left[\frac{\left(\frac{dy}{dt}\right)_{T_1}}{\left(\frac{dy}{dt}\right)_{T_2}}\right] \qquad (3)$$

For calculating activation energy, Q by Dorn method, a special heating schedule was employed. The sample was held at temperatures 1210, 1250, 1290 and 1330°C for 30 min each in case of BZY and at temperatures at 1010, 1050, 1080, 1110, 1140 and 1170°C for 30 min each for BZY with ZnO. The ramping rate between the isothermal holdings was 10°C/min.

## (4) MICROSTRUCTURE

Polished and thermally etched surfaces of a few selected samples were observed under a scanning electron microscope.

## III. RESULTS AND DISCUSSION

### (1) POWDER CHARACTERISTICS

Fig. 1 depicts the XRD patterns of the precursor powder and the powders calcined at various temperatures. The XRD pattern of the precursor powder shows the presence of $BaCO_3$. With the increase of calcination temperature, $BaCO_3$ is consumed in forming BZY which was indicated by decreasing intensity of $BaCO_3$ peaks. From this it can be inferred that phase pure material can be obtained by the coprecipitation method if the calcination is carried out at a temperature of 1100°C or above.

Particle size distribution (PSD) by laser scattering technique is shown in Fig. 2. The average particle size was found to be approximately 1 µm.

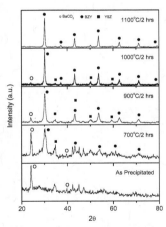

Fig. 1 : XRD Patterns of the precursor calcined at various temperatures

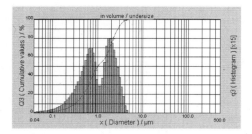

Fig. 2 : Particle size distribution curve for calcined and ground powder

## (2) COMPACTION BEHAVIOUR OF POWDERS

The semilog plot of pressure-density relation is shown in Fig. 3. Although the rate of increase of density in the lower pressure range is less, there is an appreciable increase in density with applied pressure above a certain critical value. The situation may be viewed as two straight lines separated by a break point, which is considered as a measure of agglomeration strength of the powder[8]. The strength of agglomerates found in the present investigation was well below 100 MPa. This indicates the presence of soft agglomerates in the calcined powder.

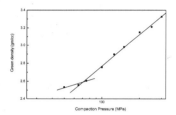

Fig. 3 : Variation of green density with applied pressure

## (3) SINTERING STUDY

(A) Dilatometric study - A representative shrinkage profile of the powder compact without ZnO and with 1 wt % ZnO is shown in Fig. 4. Difference in sintering behaviour between the two is quite evident. Table 1 lists the onset of sintering temperature and % linear shrinkage (at 1400°C) of BZY compacts with and without ZnO. The onset of sintering temperature was found to decrease upto 1 wt % ZnO addition thereafter it increases but still less than that of BZY. The % linear shrinkage was found to increase monotonically with the increasing wt % of ZnO.

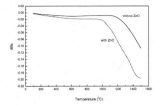

Fig. 4 : Dynamic sintering curve of Ba(Zr$_{0.85}$Y$_{0.15}$)O$_{3-\delta}$ compacts with and without ZnO

Table 1 : Onset of sintering temperature ($^{\circ}$C) and percentage linear shrinkage for BZY with various wt % of ZnO

| Wt % ZnO | Onset of sintering temperature ($^{\circ}$C) | % linear shrinkage at 1400 ($^{\circ}$C) |
|----------|-----------|-----------|
| 0 | 1379 | 1.67 |
| 0.5 | 1196 | 10.21 |
| 1 | 1070 | 14.89 |
| 2 | 1132 | 17.05 |
| 4 | 1135 | 17.39 |

(B) Densification behaviour of powder compacts – The sintered density of BZY powder compacts with various amounts of ZnO, sintered at 1300$^{\circ}$C and 1400$^{\circ}$C is given in Table 2. ZnO was found to influence the sintered density. The density was found to increase upto 1 wt % ZnO. Decrease in density was observed beyond 1 wt % ZnO addition.

Table 2 : Sintered density (%TD) of Ba(Zr$_{0.85}$Y$_{0.15}$)O$_{3-\delta}$ with different wt % of ZnO

| Sintering Temp/time | Wt % ZnO | | | | |
|---------|----|-----|----|----|----|
| | 0 | 0.5 | 1 | 2 | 4 |
| 1300$^{\circ}$C/3h | 60 | 85 | 97 | 94 | 95 |
| 1400$^{\circ}$C/3h | 72 | 91 | 98 | 93 | 95 |

(C) Sintering kinetic study – From the present study and literature[3], it has already been demonstrated that small amount (1 wt %) of zinc oxide addition is extremely effective in promoting the densification of yttria doped barium zirconate. 1 wt % of Zinc oxide can reduce the sintering temperature to below 1400$^{\circ}$C. However, the exact role of ZnO has not been understood. The sintering kinetics of of BZY powders has not been studied. In the present investigation, a comparative study of initial sintering behaviour under constant rate of heating condition between BZY and BZY with 1 wt % ZnO using two different methods has been pursued.

Plots obtained using Young & Cutler model[5] and Woolfrey-Bannister equation[6] have been shown in  Fig. 5. For BZY, both the approaches were showing a single straight line and "nQ" value calculated from their respective slopes. Q value was determined by the Dorn method[7] and n value was calculated. The value of n was found close to 0.5 indicating volume diffusion as the sintering mechanism in the investigated temperature regime. However, in presence of zinc oxide, two different segments with two different nQ values were observed. In the low temperature regime, n value was found to close to 0.5 indicating volume diffusion as the mechanism where as in the high temperature regime the value of n was calculated to be close to 1. n value close to 1 is generally observed for liquid phase sintering[4]. So, presence of ZnO was found to influence the sintering mechanism. The microstructure of BZ without and with ZnO (Fig. 6) supports this fact.

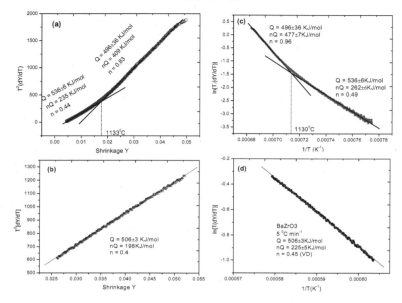

Fig. 5 : Plot of Woolfrey & Bannister model for the composition (a) BZY+ZnO and (b) BZY. Plot of Young & Cutler model for the composition (c) BZY+ZnO and (d) BZY

Without ZnO                         With ZnO

Fig. 6: Scanning electron micrographs of polished and thermally etched surfaces of BZY samples sintered at 1400°C without and with ZnO

IV. CONCLUSIONS

The present investigation has led to the following observations;

1. Phase pure yttrium doped barium zirconate has been synthesized by coprecipitation method.

2. Compaction pressure-density curve indicates the nature of powder to be softly agglomerated.

3. BZY powder was sintered to 92 % TD at 1400°C

4. Doping with ZnO influenced the densification. It was observed that different sintering mechanisms were found to operate.

REFERENCES

1. F. Boschini, B. Robertz, A. Rulmont and R. Cloots, J. Eur. Ceram. Soc. 23(2003)3035

2. H. Wang, R. Peng, X. Wu, J. Hu and C. Xia, J. Am. Ceram. Soc. 92(2009)2623

3. P. Babilo and S. M. Haile, J. Am. Ceram. Soc. 88(2005)2362

4. D. L. Johnson, J. Appl. Phys. 40(1969)192

5. W. S. Young and I. B. Cutler, J. Am. Ceram. Soc. 53(1970)659

6. J. L. Woolfrey and M. J. Bannister, J. Am. Ceram. Soc. 55(1972)390

7. R. M. German, Wiley-Interscience Publication, 1996

8. G. L. Messing, C. J. Markhoff and L. G. McCoy, Am. Ceram. Soc. Bull. 61(1982)857

# USE OF HYDROCARBON FUEL FOR MICRO TUBULAR SOFCS

Toshio Suzuki, Md. Z. Hasan, Toshiaki Yamaguchi, Yoshinobu Fujishiro, Masanobu Awano
National Institute of Advanced Industrial Science and Technology (AIST)
2266-98 Anagahora, Shimo-shidami, Moriyama-ku,
Nagoya, 463-8560, Japan

Yoshihiro Funahashi
Fine Ceramics Research Association (FCRA)
2266-99 Anagahora, Shimo-shidami, Moriyama-ku,
Nagoya, 463-8561, Japan

Nigel. Sammes
Colorado School of Mine
1500 Illinois Street,
Golden, Colorado 80401 USA

## ABSTRACT

The feasibility of use of hydrocarbon fuel has been explored for micro tubular solid oxide fuel cells (SOFCs) consisted of Gd doped ceria (GDC) or Sc stabilized zirconia (ScSZ) electrolyte. Methane is selected for the fuel and directly supplied to the cell with water vapor. Performance of the cells with methane-water has been evaluated, and the GDC cell has showed the power density of 0.28 $Wcm^{-2}$ at 600 °C, while the ScSZ cell showed 0.45 $Wcm^{-2}$ at 650 °C furnace temperature. Both results are similar to the performance obtained with hydrogen fuel.

## INTRODUCTION

Solid oxide fuel cells (SOFCs), which display the highest energy efficiency among other types of fuel cells, have been intensively studied over 20 years all over the world [1, 2]. SOFCs can also offer high fuel flexibility, which allows using various hydrocarbon fuels with or without simple pre reformer [3-5]. Up to the present, SOFC systems have been mainly developed for stationary power plants needed large auxiliaries such as heating units and high heat-insulating packages. In addition, SOFC stacks were not durable for rapid heat cycling due to consisting of ceramic components [6, 7]. Thus, the degradation mechanism of rapid heat cycling has been intensively studied in order to improve the reliability of the SOFCs [8-10].

In order to broaden the application use of SOFCs, SOFCs operable at low operating temperature with quick start-up/shut-down ability are longed to be realized [11, 12]. So far, it was shown that small-scale tubular SOFCs endured thermal stress caused by rapid start-up operation. The micro tubular design was also suggested to realize a stack with high volumetric power density, which makes them attractive for use in many applications including auxiliary power units and potable power devices.

In our previous study, we have reported that high performance micro tubular SOFCs could be obtained by optimizing electrolyte thickness and the microstructure of electrodes, resulted in over 1 $Wcm^{-2}$ at 550 °C using ceria based electrolyte and at 600 °C using zirconia based electrolyte [13, 14].

In this study, the feasibility of use of hydrocarbon fuel has been explored for micro tubular solid oxide fuel cells (SOFCs) consisted of Gd doped ceria (GDC) or Sc stabilized zirconia (ScSZ) electrolyte. Methane was selected and used for evaluating cell performance of micro tubular SOFC's at the intermediate operating temperature below 650°C, for exploring the possibility of use of the cell under direct fuel reforming.

EXPERIMENTAL

Two types of micro tubular SOFCs were tested as shown in Fig. 1 and Table I. GDC, LSCF, and ScSZ stand for Gd doped $CeO_2$, $La_{0.6}Sr_{0.4}Co_{0.2}Fe_{0.8}O_{3-y}$, and Sc stabilized $ZrO_2$, respectively.

Anode support tubes were prepared from NiO powder (Sumitomo co., ltd.), $Gd_{0.2}Ce_{0.8}O_{2-x}$ (GDC) (Shin-Etsu Chemical co., ltd.) or $Sc_{0.1}Zr_{0.89}Ce_{0.01}O_2$ (ScSZ) (Diichikigenso co., ltd), a pore former (poly methyl methacrylate beads (PMMA), Sekisui Plastics co., ltd.), and a binder (cellulose, Yuken Kogyo co., ltd.). These powders were mixed for 1 h by a mixer 5DMV-rr (Dalton co., ltd.). After adding a proper amount of water, the mass was stirred for 30 min in a vacuumed chamber, and then, was left over 15 h for aging. The mass was used for extruding the anode tubes from a metal mold using a piston cylinder type extruder (Ishikawa-Toki Tekko-sho co., ltd.).

A slurry for dip-coating electrolyte was prepared by mixing the GDC powder or the ScSZ powder, solvents (methyl ethyl ketone and ethanol), a binder (poly vinyl butyral), a dispersant (polymer of an amine system) and a plasticizer (dioctyl phthalate) for 24 h. The anode tubes were dip-coated using the slurry at the pulling rate of 1.5 mm/sec. The tubes with coating layer were dried in air, and sintered at 1300°C for 1 h in air. The anode tubes with electrolyte were, again, dip-coated using a cathode slurry, which was prepared in the same manner using $La_{0.6}Sr_{0.4}Co_{0.2}Fe_{0.8}O_{3-y}$ (LSCF) powder (Seimi Chemical, co., ltd.), the GDC powder, and organic ingredients. After dip-coated, the tubes were dried and sintered at 1050°C for 1 h in air to complete a cell. In the case of the ScSZ cell, GDC slurry prepared by the same manner was dip-coated on the electrolyte and sintered at 1200°C before cathode coating.

The cell performance was investigated by using a using a Parstat 2273 (Princeton Applied Research) in DC 4 point probe measurement. The cell size was 1.8 mm in diameter and 30 mm in length with cathode length of 22 mm, whose effective cell area was 1.24 $cm^2$. Note that these cells were designed to use for the actual for 50-200W class stack/modules [15]. The Ag wire was used for collecting current from anode and cathode sides, which were both fixed by Ag paste. The current from anode side was collected from an edge of the tube. After the application of the current collector, the cell was fixed in the an alumina sample holder by using a ceramic bond. The sample holder was then, set in the furnace and mixture of methane, steam with nitrogen (20/20/25 mLmin$^{-1}$) was flowed inside of the tubular cell. The air of 100 mLmin$^{-1}$ was flowed at the cathode side. The cell temperature was measured using a thermocouple placed at the middle position of the tubular cell.

Table I: Structure of micro tubular SOFCs tested in this study

| cell type | anode | electrolyte | interlayer | cathode |
|-----------|-------|-------------|------------|---------|
| GDC cell | Ni-GDC | GDC | - | LSCF-GDC |
| ScSZ cell | Ni-ScSZ | ScSZ | GDC | LSCF-GDC |

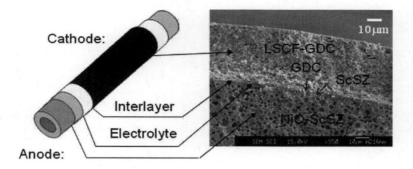

Figure 1: Appearance of micro tubular SOFC (SEM image: ScSZ cell).

RESULTS AND DISCUSSION
(a) GDC cell
    Figures 2 show the performance of the GDC cell obtained using hydrogen fuel and the mixture gas ($CH_4 + H_2O + N_2$) at the furnace temperature of $600°C$ (cell temperature $\sim550°C$). The open circuit voltage of about 0.9 V was obtained for both hydrogen fuel and the mixture gas, which indicated that the anode has sufficient catalytic activity for steam reforming below 600 °C. As can be seen in Fig. 2 (a), the maximum power densities of the cell for $H_2$ fuel and $CH_4 + H_2O$ fuel are almost identical, reached $0.28 Wcm^{-2}$. The cell was stably operated under the current load of $0.16\ Acm^{-2}$ using $CH_4 + H_2O$ fuel as shown in Fig. 2(b). Thus, it was shown that direct steam reforming operation can be possible using Ni-GDC anode under $600°C$ (cell temperature).

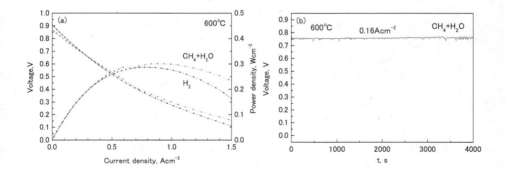

Figure 2: Performance of the GDC cell (a) I-V characterization of the cell using $H_2$ and $CH_4 + H_2O$ fuel (b) The results of current load test of $0.16\ Acm^{-2}$ at $600°C$ furnace temperature.

(b) ScSZ cell

Figures 3 show the results of (a) I-V characterization and (b) impedance measurement for the ScSZ cell using $H_2$ fuel obtained at 600~700 °C furnace temperature (cell temperature 550~650 °C). As can be seen, the maximum power densities of the cell were 0.2, 0.45 and 0.6 Wcm$^{-2}$ for the furnace temperature at 600, 650 and 700 °C, respectively. Each impedance spectrum in Fig. 3(b) showed two semi-circles, which could be correlated to the effect of charge transfer (chemical reactions) and gas transport (diffusion). It is indicated that the low frequency semi-circle was effectively reduced by increasing temperature from 600 to 650 °C.

The results of direct reforming operation using $CH_4+H_2O$ fuel were shown in Figs. 4 obtained at 650 °C furnace temperature. As previously shown in the GDC cell part, the ScSZ cell operated at direct methane steam reforming condition also showed similar performance obtained with $H_2$ fuel. The cell was also operated under the current load of 0.16 Acm$^{-2}$ as shown in Fig. 4 (b) for a few hours, as an initial check, which turned out to be stable.

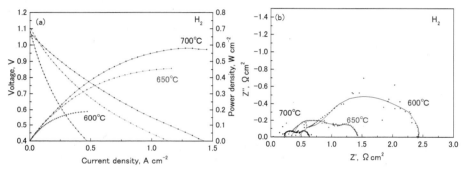

Figure 3: Performance of the ScSZ cell using $H_2$ fuel (a) I-V characterization and (b) impedance analysis.

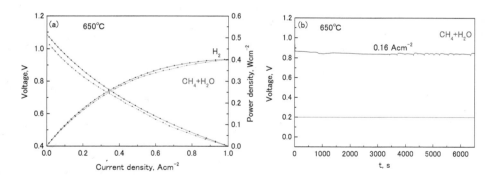

Figure 4: Performance of the ScSZ cell (a) I-V characterization of the cell using $H_2$ and $CH_4+H_2O$ fuel (b) Current load test of 0.16 Acm$^{-2}$ at 600°C furnace temperature.

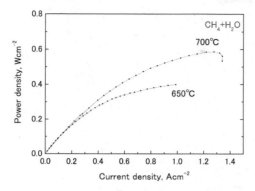

Figure 5: I-V characterization of the cell using $CH_4+H_2O$ fuel at various temperatures.

Actually the ScSZ cell can be operated at higher temperatures, and the result of power characterization obtained at higher temperature was shown in Fig. 5 using $CH_4+H_2O$ fuel. The maximum power density of 0.6 $Wcm^{-2}$ was obtained at 700 °C furnace temperature (cell temperature ~ 650 °C). Currently, long term test is undergoing using the GDC cell and the ScSZ cell under direct reforming conditions.

CONCLUSION

In this study, the cell performance under the direct reforming operation of methane was evaluated using the cells with ceria based electrolyte and zirconia based electrolyte. The maximum power densities of the GDC cell for $H_2$ fuel and $CH_4+H_2O$ fuel are almost identical, reached $0.28Wcm^{-2}$. The cell was stably operated under the current load of 0.16 $Acm^{-2}$ using $CH_4+H_2O$ fuel at 0.75V. It was confirmed that direct steam reforming operation can be possible under 600°C using the GDC cell. The maximum power density of the ScSZ cell using $CH_4+H_2O$ fuel was 0.4 and 0.6 $Wcm^{-2}$ for the furnace temperature at 650 and 700 °C, respectively (cell temperature 600, 650 °C). The ScSZ cell also showed similar performance to $H_2$ fuel. The ScSZ cell was also stably operated under the current load of 0.16 $Acm^{-2}$ at 0.85 V.Currently, long term operation of the cells using $CH_4+H_2O$ fuel was conducted for practical application use.

ACKNOWLEDGEMENT
This work is supported by NEDO, as part of the Advanced Ceramic Reactor Project.

REFERENCES
[1] O.Yamamoto, *Electrochimica Acta*, 45 (15), pp.2423-2435 (2000).
[2] S. C. Singhal, Solid State Ionics, 152, pp.405 – 410 (2002).

[3] H. Sumi, K. Ukai, Y. Mizutani, H. Mori, C-J. Wen, H. Takahashi, O. Yamamoto, *Solid State Ionics*, 174 (1), pp.151 – 156 (2004)

[4] T. Matsui, T. Iida, M. Kawano, T. Inagaki, R. Kikuchi, K. Eguchi, *ECS Trans.*, 7 (1), pp.1437 (2007)

[5] K. Sasaki, K. Susuki, A. Iyoshi, M. Uchimura, N. Imamura, H. Kusada, Y. Teraoka, H. Fuchino, K. Tsujimoto, Y. Uchida, N. Jingo, *J. Electrochem. Soc.*, 153 (11), pp.A2023 - A2029 (2006)

[6] Y. Mizutani, K. Hisada, K. Ukai, H. Sumi, M. Yokoyama, Y. Nakamura, O. Yamamoto, *J. Alloys and Compounds*, 408, pp.518 – 524 (2006)

[7] T. Inagaki, F. Nishiwaki, S. Yamasaki, T. Akbay, K. Hosoi, *J. Power Sources,* 181 (2), pp.274 – 280 (2008)

[8] D. Waldbillig, A. Wood, D.G.Ivey, *Solid State Ionics*, 176 (9), pp.847 – 859 (2005)

[9] A. Hagen, Y. L. Liu, R. Barfod, P. V. Hendriksen, *J. Electrochem. Soc.*, 155 (10), pp. B1047 – B1052 (2008)

[10] K. Eguchi, Y. Kunisa, K. Adachi and H. Arai, *J. Electrochem. Soc.,* 143 (11), pp.3699 – 3703 (1996)

[11] W. Bujalski, C. M. Dikwal and K. Kendall, *J. Power Sources*, 171 (1), pp.96 – 100 (2007)

[12] K. Kendall, C. M. Dikwal and W. Bujalski, *ECS Trans.*, 7 (1), pp.1521 – 1526 (2007)

[13] T. Suzuki, Y. Funahashi, T. Yamaguchi, Y. Fujishiro, M. Awano. *Electrochem. Solid State Lett.*10-8 pp.177-179 (2007)

[14] T. Suzuki, Z. Hasan, Y. Funahashi, T. Yamaguchi, Y. Fujishiro, M. Awano. *Science* 325, pp.852-855 (2009)

[15] Y. Funahashi, T. Shimamori, T. Suzuki,Y. Fujishiro and M. Awano, *ECS Trans.*, 25-2, pp.195-200 (2009)

# PHASE DIAGRAM OF PROTON-CONDUCTING Ba(Zr$_{0.8-x}$Ce$_x$Y$_{0.2}$)O$_{2.9}$ CERAMICS BY *IN SITU* MICRO-RAMAN SCATTERING AND X-RAY DIFFRACTION

C.-S. Tu[a,b], C.-C. Huang[b], S.-C. Lee[b], R. R. Chien[c], V. H. Schmidt[c], and J. Liang[b]
[a]Graduate Institute of Applied Science and Engineering, Fu Jen Catholic University, Taipei 242, Taiwan, Republic of China
[b]Department of Physics, Fu Jen Catholic University, Taipei, Taiwan 242, Republic of China
[c]Department of Physics, Montana State University, Bozeman, MT 59717, USA

ABSTRACT

*In situ* X-ray diffraction (XRD) and micro-Raman scattering have been used to study structures, lattice parameters, Raman vibrations, and phase transitions of Ba(Zr$_{0.8-x}$Ce$_x$Y$_{0.2}$)O$_{2.9}$ (BZCY) ceramics ($x$ = 0.0-0.8) as a function of temperature (20-900 °C). The glycine-nitrate (G/N) combustion process was used to synthesize nano-sized proton-conducting BZCY ceramic powders as a function of cerium-to-zirconium ratio. The glycine-to-nitrate molar ratio used to fabricate BZCYs is G/N = 1/2. Second phase was not observed on BZCY ceramics after calcining at 1400 °C for 5 hours. A first-order structure transition from rhombohedral (R) to cubic (C) takes place upon heating for $x$ = 0.5-0.8, in which transition temperature shifts toward higher temperature as cerium content increases, except for BCY82 ($x$ = 0.8) which has a more complex transition sequence. The R–C phase transition was not observed for lower-cerium BZCYs ($x$ = 0.0-0.4) in the region of 20-900 °C, indicating a cubic phase at and above room temperature.

INTRODUCTION

Perovskite-type (Ba,Sr,Ca)(Zr,Ce)O$_3$ oxides exhibit good protonic conduction under hydrogen-containing atmosphere at elevated temperature, and are promising for applications of proton-conducting solid oxide fuel cells (H-SOFCs), hydrogen separation membranes, hydrogen pumps, hydrogen sensors, and steam electrolyzers for hydrogen production.[1-3] For instance, hydrogen-fueled H-SOFCs with BZCY electrolytes made by Yang *et al.*[4] achieved power density of 700 mW/cm$^2$ at the relatively low temperature of 700 °C, and BZCY-based cells made by Meng *et al.*[5] reached 371 mW/cm$^2$. Our group is obtaining comparable power densities with BZCY-based cells, and this work was submitted for publication. These power densities are competitive with those achieved by O-SOFCs with oxygen ion conducting electrolytes. H-SOFCs have two advantages over O-SOFCs, based on the fact that on the anode side there is only hydrogen inflow, with no steam outflow. First, higher fuel utilization can be achieved because there is no loss of fuel flowing out with the steam exhaust gas. Second, anode supported cells having a relatively thick anode are desirable because the anode has a strong metal (usually Ni) framework, and there is much less fuel pressure loss through the anode passages in H-SOFCs because there is no steam counterflow.

Cerate-based proton conductors have a high ionic conductivity but exhibit poor chemical stability in CO$_2$ and H$_2$O containing atmosphere at elevated temperature.[6,7] Although zirconate-based proton conductors have lower ionic conductivity, they show good chemical and mechanical stabilities.[8] These results suggest that proton-conducting solid solutions between cerate and zirconate may have both high proton conductivity and good chemical stability.[9,10] A$^{II}$B$^{IV}$O$_3$-based proton conductors are doped in B-site by lower valence elements, typically Y$^{3+}$ (R$^{III}$ = 0.9 Å) or trivalent rare earth metal cations, creating oxygen vacancies. Subsequent exposure to humid atmospheres is presumed to lead to the incorporation of protons, resulting in proton conduction.[11,12]

BaCeO$_3$ has a phase sequence of orthorhombic (*Pnma*)–orthorhombic (*Imma*)–rhombohedral ($R\bar{3}c$)–cubic ($Pm\bar{3}m$) at 290, 400, and 900 °C.[13]  Melekh *et al.*[14] reported that different workers found orthorhombic, cubic, pseudocubic, and tetragonal phases at room temperature for BaCeO$_3$ prepared by various methods.  This type of perovskite can crystallize in various phases, depending on processing method and sintering temperature.  BaZrO$_3$ is cubic at and above room temperature.[15]  Neutron powder diffraction measurement suggested that Ba(Ce$_{0.8}$Y$_{0.2}$)O$_{2.9}$ (BCY82) has a rhombohedral $R\bar{3}c$ phase at room temperature.[16]  Neutron scattering result proposed a coexistence of rhombohedral $R\bar{3}c$ and monoclinic *12/m* phases at room temperature under different atmospheric conditions.[17]

There is not much knowledge of the temperature-dependent phase diagram regarding the proton-conducting Ba(Zr,Ce,Y)O$_{3-\delta}$ ceramics.  In this work, *in situ* temperature-dependent XRD and micro-Raman were employed to investigate structure transitions and the phase diagram of calcined Ba(Zr$_{0.8-x}$Ce$_x$Y$_{0.2}$)O$_{2.9}$ powders in a argon-containing environment in the region of 20-900 °C.

## EXPERIMENTAL

The glycine-to-nitrate (G/N) process (with G/N=1/2) was used to synthesize nano-sized Ba(Zr$_{0.8-x}$Ce$_x$Y$_{0.2}$)O$_{2.9}$ (BZCY) ceramic powders as a function of cerium-to-zirconium ratio ($x$ = 0.0-0.8).  All synthesized powders were calcined at 1400 °C for 5 hours in the laboratory air.  For *in situ* X-ray diffraction measurements, a high-temperature Rigaku Model MultiFlex X-ray diffractometer with Cu K$\alpha_1$ ($\lambda$ = 0.15406 nm) and Cu K$\alpha_2$ ($\lambda$ = 0.15444 nm) radiations was used as a function of temperature (20-900 °C) in argon air.  The calcined BZCYs powders were placed and smoothly pressed on the platinum sample holder.  The reason for using argon air as a working atmosphere is to avoid possible reaction with CO$_2$ in the air.

For micro-Raman scattering, a double grating Jobin Yvon Model U-1000 double monochromator with 1800 grooves/mm gratings and a nitrogen-cooled CCD as a detector were employed.  A Coherent Model Innova 90 argon laser of wavelength $\lambda$ = 514.5 nm was used as an excitation source and the Raman spectra were collected in the region of 150-1000 cm$^{-1}$ as a function of temperature.

## RESULTS AND DISCUSSION

Figure 1 shows the XRD and micro-Raman spectra of BZCYs ($x$ = 0.0-0.8) ceramic powders at room temperature.  A second phase was not observed in all compounds, indicating that the optimal glycine-to-nitrate process (G/N=1/2) and calcining conditions can fabricate the single-phase powders.  The main reflection peaks of BZY82 include (100), (110), (111), (200), (211), (220), (310), and (222), suggesting a simple-centered cubic (sc) unit cell according to the structure-factor calculation.[18]  The cubic lattice parameter was estimated from the (110) peak ($2\theta$ = 29.75°) and is $a$ = 4.2436 Å for BZY82 at room temperature.  As indicated by the dashed line, the XRD peak shifts toward lower $2\theta$ degrees as cerium content increases, because Ce$^{4+}$ (R$^{VI}$= 0.87 Å) has a larger ionic radius than Zr$^{4+}$ (R$^{VI}$= 0.72 Å).[19]  The XRD spectra show a two-peaks splitting for higher-cerium BZCYs ($x$ =0.5-0.8), implying a different structure (perhaps rhombohedral phase) at room temperature.  Note that neutron powder diffraction suggested that Ba(Ce$_{0.8}$Y$_{0.2}$)O$_{2.9}$ (BCY82) has an rhombohedral $R\bar{3}c$ phase at room temperature.[16]  A coexistence of rhombohedral $R\bar{3}c$ and monoclinic *12/m* phases at room temperature under different atmospheric condition was also proposed from neutron scattering.[17]

The major vibrations of Raman modes of BZY82 appear near 240, 370, 420, 500, 560, and 720 cm$^{-1}$.  As indicated by the dashed lines, vibration modes shift toward lower frequency as cerium content increases, due to the larger atomic mass of cerium compared with Zr. As indicated by the "*", a broad Raman mode as a shoulder begins to appear near 380 cm$^{-1}$ in BZCY442, and its intensity grows stronger as cerium increases. This is consistent with the XRD result, in which an apparent two-peaks splitting occurs in the higher-cerium BZCYs ($x$ = 0.5-0.8).

The major Raman vibrations of Y$_2$O$_3$, CeO$_2$, and ZrO$_2$ are 375, 461, and 474 cm$^{-1}$, respectively. The 461 cm$^{-1}$ vibration of CeO$_2$ results from the F$_{2g}$ Raman-active mode of fluorite structure.[20] The 474 cm$^{-1}$ of ZrO$_2$ corresponds to the A$_g$ Raman-active mode of O-O vibration.[21] The Raman modes near 370 cm$^{-1}$ likely associates with Y$_2$O$_3$-like structure, which is sensitive to the change of cerium content. The vibrations appear in the region 440-500 cm$^{-1}$ likely associate with (Zr,Ce,Y)O$_2$ structure.

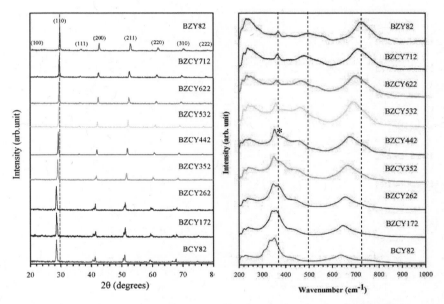

Figure 1 (a) X-ray diffraction and (b) micro-Raman spectra of BZCY ($x$ = 0.0-0.8) ceramic powders taken at room temperature. The dashed lines are guides for shifts of 2θ or vibration frequency as cerium content increases.

To see the development of phase transformation as a function of cerium content, the (110) XRD spectra of BZCY ($x$ = 0.5-0.8) powders taken at room temperature are enlarged in Figure 2. Solid and dashed lines represent reflections from Kα$_1$ and Kα$_2$ radiations, respectively. The XRD spectra were fitted by using the PeakFit software with the sum of Gaussian and Lorentzian terms. The intensity ratio between Kα$_1$ and Kα$_2$ radiations is about 2:1.[18] A single XRD peak (from each radiation) appears in the lower-cerium compounds ($x$ ≤ 0.4), indicating the cubic phase. As cerium increases, two XRD peaks (from each radiation) were observed and imply a rhombohedral phase at room temperature for $x > 0.4$.

Figure 3 shows temperature-dependent Raman vibration modes (near 630-730 cm$^{-1}$) and $d$ spacings calculated from the (110) XRD peak of BZCYs powders upon heating for the higher-cerium compounds ($x = 0.5$-$0.8$). The 2θ-reflection and $d$ spacing of XRD obey the Bragg law, i.e. $2d_{hkl}\sin\theta_{hkl} = n\lambda$. Two $d$ spacings are expected from the (110) reflections for rhombohedral structure, i.e, $1/d^2 = [(h^2+k^2+l^2)\sin^2\alpha + 2(hk+kl+hl)(\cos^2\alpha - \cos\alpha)]/[a_R^2(1-3\cos^2\alpha + 2\cos^3\alpha)]$, where $(h,k,l)$ and $(a_R, \alpha)$ are crystallographic orientation and lattice parameters of rhombohedral unit cell, respectively.[18]

The Raman vibration modes (near 630-730 cm$^{-1}$) of higher-cerium BZCYs ($x = 0.5$-$0.8$) exhibit a step-like increase as temperature approaches the structural transition, while two-peaks $d$ spacings merges into one $d$ spacing. The transition temperature shifts toward higher temperature as cerium content increases, except for BCY82. This phase transition from rhombohedral to cubic phase is likely first-order.

Figure 4 shows temperature-dependent lattice parameters ($a$ and $\alpha$) calculated from the (110) XRD peaks of BZCY ($x = 0.5$-$0.8$) powders upon heating. The lattice parameters exhibit a step-like anomaly as temperature approaches the transition temperature as indicated by the dashed line. The gradual increase of lattice parameters is mainly due to thermal expansion of unit cell. Results for lower-cerium BZCY (x = 0.0-0.4) are not shown because they have no obvious phase transition above room temperature. The phase diagram (transition temperature vs. Ce content) of BZCY ($x = 0.0$-$0.8$) is summarized in Fig. 5.

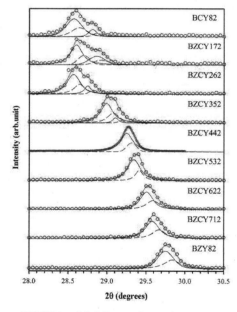

Figure 2 (110) XRD spectra of BZCY ($x = 0.5$-$0.8$) ceramic powders at room temperature. Solid and dashed lines represent reflections from Kα$_1$ and Kα$_2$ radiations, respectively.

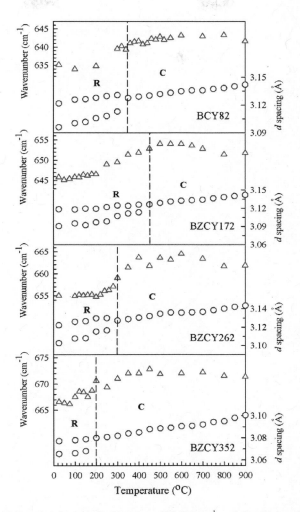

Figure 3  Temperature-dependent Raman vibration (near 630-730 cm$^{-1}$) and $d$ spacings calculated from the (110) XRD peak of BZCY ($x$ = 0.5-0.8) ceramic powders upon heating.  The dashed lines indicate transition temperatures.  R and C represent rhombohedral and cubic phases, respectively.

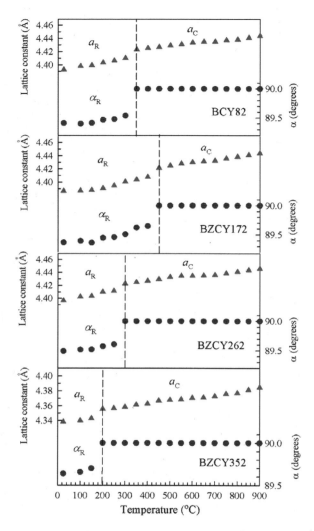

Figure 4  Temperature-dependent lattice parameters calculated from the (110) XRD peak of BZCY ($x$ = 0.5-0.8) ceramic powders upon heating.  The dashed lines indicate the R-C transition temperatures.

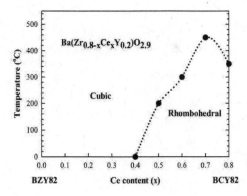

Figure 5 Phase diagram (transition temperature vs. Ce content) of calcined BZCY ($x$ = 0.0-0.8) ceramic powders obtained from the heating process. The dotted line is a guide for eyes.

CONCLUSIONS

*In situ* temperature-dependent X-ray diffraction and micro-Raman scattering have been used to investigate structures, lattice parameters, atomic vibration modes, and phase transitions of calcined BZCY ceramics ($x$ = 0.0-0.8) synthesized by the glycine-nitrate combustion process with G/N = 1/2. A second phase was not observed for BZCY powders calcined at 1400 °C for 5 hours. A rhombohedral – cubic phase transition was observed in higher cerium compounds ($x$ = 0.5-0.8) and the transition shifts toward higher temperature as cerium content increases, except for BCY82. The R–C phase transition was not observed for $x$ = 0.0-0.4 in the region of 20-900 °C, indicate a cubic phase at and above room temperature.

ACKNOWLEDGEMENT

This work was supported by National Science Council of Taiwan Grant No. 96-2112-M-030-001, and by DOE under subcontract DE-AC06-76RL01830 from Battelle Memorial Institute and PNNL.

REFERENCES

1. H. Iwahara, T. Esaka, H. Uchida, N. Maeda, Solid State Ionics 3/4 (1981) 359-363.
2. H. Iwahara, H. Uchida, K. Kondo, K. Ogaki, J. Electrochem. Soc. 135 (1988) 529-533.
3. H. Iwahara, Solid State Ionics 77 (1995) 289-298.
4. L. Yang, C. Zuo, S. Wang, Z. Cheng, M. Liu, Adv. Mater. 20, (2008) 3280-3283.
5. G. Meng, G. Ma, Q. Ma, R. Peng, X. Liu, Solid State Ionics 178 (2007) 697-703.
6. N. Bonanos, K.S. Knight, B. Ellis, Solid State Ionics 79 (1995) 161-170.
7. C.W. Tanner, A.V. Virkar, J. Electrochem. Soc. 143 (1996) 1386-1389.
8. K.D. Kreuer, Solid State Ionics 97 (1997) 1-15.
9. C. Zuo, S. Zha, M. Liu, M. Hatano, M. Uchiyama, Adv. Mater. 18 (2006) 3318-3320.

10. Z. Zhong, Solid State Ionics 178 (2007) 213-220.
11. T. Norby, M. Widerøe, R. Glöckner, Y. Larring, Dalton Trans., (2004) 3012-3018.
12. K.D. Kreuer, Annu. Rev. Mater. Res. 33 (2003) 333-359.
13. K. S. Knight, Solid State Ionics 145, (2001) 275-294.
14. B.-T. Melekh, V. M. Egorov, Y. M. Baikov, N.F. Kartenko, Y. N. Filin, M. E. Kompan, I. I. Novak, G. B. Venus, V. B. Kulik, Solid State Ionics 97 (1997) 465-470.
15. I. Charrier-Cougoulic, T. Pagnier, G. Lucazeau, J. Solid State Chem. 142 (1999) 220-227.
16. K. Takeuchia, C.-K. Loong, J. W. Richardson Jr., J. Guanb, S.E. Dorris, U. Balachandran, Solid State Ionics 138 (2000) 63–77.
17. C.-K. Loong, M. Ozawa, K. Takeuchi, K. Ui, N. Koura, J. Alloys and Compounds 408–412 (2006) 1065–1070.
18. B.D. Cullity, Elements of X–ray diffraction, (Addison–Wesley Publishing, 1978).
19. R. D. Shannon, Acta Crystallographia. A32 (1976) 751-767.
20. R. Q. Long, Y. P. Huang, H. L. Wan, J. Raman Spectroscopy 28 (1997) 29-32.
21. B.-K. Kim, H.-O. Hamaguchi, Phys. Stat. Sol. (b) 203 (1997) 557-563.

# ELECTRICAL CONDUCTIVITY OF COMPOSITE ELECTROLYTES BASED ON BaO-CeO$_2$-GdO$_{1.5}$ SYSTEM IN DIFFERENT ATMOSPHERES

A. Venkatasubramanian, P. Gopalan and T.R.S. Prasanna
Department of Metallurgical Engineering and Materials Science
Indian Institute of Technology, Bombay
Mumbai, Maharashtra, India

ABSTRACT
    Different methods have been employed to reduce the electronic conductivity of Gd doped ceria for potential application as an electrolyte in intermediate temperature solid oxide fuel cell. In the present study, addition of proton/oxygen ion conducting Gd doped barium cerate to Gd doped ceria has been investigated. Composites with nominal compositions Ce$_{0.8}$Gd$_{0.2}$O$_{1.9}$ + x BaO (0.1 ≤ x ≤ 0.4) have been synthesized through the conventional solid state reaction route. The X-ray diffraction (XRD) studies reveals presence of two phases in all composites. One phase matches with Fluorite, Ce$_y$Gd$_{(1-y)}$O$_{2-(y/2)}$ (CG) and other one matches with Pervoskite BaCe$_z$Gd$_{(1-z)}$O$_{3-(z/2)}$ (BCG). Electrical conductivity measurements of all composite electrolytes have been carried out in dry nitrogen, H$_2$O and D$_2$O-saturated atmosphere between 100$^{\circ}$C and 600$^{\circ}$C. In the temperature range from 160$^{\circ}$C to 260$^{\circ}$C, the activation energy for the bulk conductivity of the composite electrolytes at x ≥ 0.2 show an observable change in different atmospheres, indicating presence of proton conduction. In the temperature range between 350$^{\circ}$C and 600$^{\circ}$C, the activation energy for the total conductivity of the composite electrolyte at x=0.4 shows significant changes in different atmospheres.

## INTRODUCTION

    A Solid Oxide Fuel Cell (SOFC) can convert chemical energy into electrical energy without any intermediate step. An electrolyte is one of the critical components of a SOFC; plays a significant role in deciding power density at a given operating temperature[1,2]. The state-of-the-art SOFCs use yttria stabilized zirconia (YSZ) based electrolytes[3]. However, owing to the high resistivity of YSZ; these SOFCs have to operate around 1000$^{\circ}$C[4]. Finding electrolytes with sufficient ionic conductivity around 800$^{\circ}$C can considerably reduce the cost of balance of plant components. Gd-doped ceria[5,6] is one of the candidate electrolyte materials for lowering operating temperature of SOFCs. In this material vacancies are formed due to the introduction of Gd into the CeO$_2$ lattice as shown in Eq. (1). However, these materials exhibit electronic conduction at high temperature and low partial pressures of oxygen present at the anode side of a SOFC[7].

$$Gd_2O_3 \rightarrow 2Gd_{Ce}^{'} + V_O^{\bullet\bullet} + 3O_o^X \tag{1}$$

Introduction of Gd into the BaCeO$_3$ lattice results in the formation of oxygen vacancies as shown by Eq. (2). When this material is exposed to atmosphere containing H$_2$O; protonic defects are introduced into the BaCeO$_3$ lattice as shown in Eq. (3). In doped BaCeO$_3$, oxygen ions and protons are the charge carriers depending upon the temperature and atmosphere[8,9]. In these materials protons and oxygen vacancies co-exist; at low temperatures protonic contribution to the total conductivity is significant[10,11].

$$2Ce_{Ce}^X + O_o^X + Gd_2O_3 \rightarrow 2Gd_{Ce}^{'} + V_O^{\bullet\bullet} + 2CeO_2 \tag{2}$$

$$H_2O + V_O^{\bullet\bullet} + O_o^x \leftrightarrow 2OH_o^{\bullet} \tag{3}$$

Protonic defects migrate through the fixed oxygen ion sub-lattice without association with any particular oxygen ion[12]. In this material protonic conduction can be confirmed by isotope effect. Deuterium ions carry the same charge as that of proton, but their mass is almost double that of proton. When deuterium ions are present in the lattice instead of protons; results in an observable change in conductivity. This serves as a confirmatory test for protonic conduction. Classical theory[13] predicts the pre-exponential term of proton conductivity is higher than that of deuteron conductivity by a factor of $\sqrt{2}$. On the other hand, quantum theory[13] based calculations predict higher activation energy for deuterium ion conduction than proton conduction (around 0.06 eV). Haile et al.[14] and Stevenson et al.[15] have measured conductivity of BaCe$_{0.85}$Gd$_{0.15}$O$_{3-\delta}$ in H$_2$O and D$_2$O-saturated atmospheres and confirmed the protonic conductivity. The variation in conductivity is consistent with the trends predicted by the theories described earlier.

By incorporating Gd-doped barium cerate into Gd-doped ceria matrix; proton conductivity can be introduced into composites. Consequently, these composites are likely to exhibit isotope effect in D$_2$O-saturated atmosphere. Impedance spectroscopy studies carried out in dry nitrogen, H$_2$O and D$_2$O-saturated atmospheres can give insight into the nature of grain, grain boundary and interface effects in the composites.

EXPERIMENTAL DETAILS

Stoichiometric amount of BaCO$_3$ (99.95%, Alfa Aesar, USA), CeO$_2$ (99.9%, Sigma Aldrich, USA) and Gd$_2$O$_3$ (99.9%, Alfa Aesar, USA) for powder compositions listed in Table 1 were ball milled and calcined twice at 1350°C for 6 h. Green pellets from powders were obtained by uniaxial pressing at 100 MPa, using a steel die of 10 mm diameter. The pellets were sintered at 1550°C for 10 h in air. Samples for impedance measurements listed in Table I, were prepared by the method described in a previous work[16]. A Hewlett Packard HP4192A LF impedance analyzer was used for the impedance measurements. In the temperature range 100°C-600°C, the impedance measurements were carried out over the frequency range $10^2$-$10^7$ Hz; with an applied potential of 1 V. Data were collected in the cooling cycle. All impedance measurements in different atmospheres were carried out in a Probostat$^{TM}$ sample holder[17]. The samples were equilibrated at 600°C for 6 h in dry N$_2$ or N$_2$ saturated with H$_2$O or N$_2$ saturated with D$_2$O before the conductivity measurements. Impedance data were analyzed using Zview software[18]. The impedance data were fitted to an equivalent circuit containing resistance – constant phase element (R-CPE) circuits using the Zview software. The capacitance values were extracted using the formula

$$C = Y^{(1/n)}R^{(1/n-1)} \tag{4}$$

where, R being resistance, Y and n being two parameters that characterize CPE used for fitting. Temperature dependent conductivity values were plotted and fitted to the following equation

$$\sigma = \frac{\sigma_0}{T}\exp\left(-\frac{E_a}{kT}\right) \tag{5}$$

where, $\sigma$ is the conductivity, $\sigma_0$ is a pre-exponential term, T is the absolute temperature, $k$ is Boltzmann constant and $E_a$ is the activation energy for conduction.

Table I. Compositions for the present investigation.

| Sample code | Ce$_{0.8}$Gd$_{0.2}$O$_{1.9}$ + x BaO |
|---|---|
| Ba10CG | x=0.1 |
| Ba20CG | x=0.2 |
| Ba30CG | x=0.3 |
| Ba40CG | x=0.4 |

## RESULTS AND DISCUSSION

XRD patterns of the composite powders calcined at 1350°C are shown in Figure 1. The XRD patterns show presence of two phases. One phase is identified as Ce$_y$Gd$_{(1-y)}$O$_{2-(y/2)}$ (CG) and other one as BaCe$_z$Gd$_{(1-z)}$O$_{3-(z/2)}$ (BCG). The relative intensity of BCG phase increases with the BaO content. The back scattered image of sample Ba40CG shown in Figure 2. This image show bright and dark regions corresponding to CG and BCG phase, respectively. The average grain size of BCG grains is higher than that of CG grains. The lines running parallel in the SEM image are likely due to the polishing of samples with 400 grit SiC paper, which has been used to remove the platinum coating after the conductivity measurements.

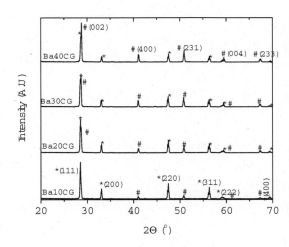

Figure 1. XRD pattern of composite powders calcined at 1350°C.(* and # represent CG and BCG phase, respectively)

Figure 2. Back scattered electron image of sintered Ba40CG.

Figure 3 shows typical impedance spectra of sample Ba20CG at 200°C in different atmospheres. These impedance spectra show two over lapped semi-circles; henceforth known as high frequency (HF) and intermediate frequency (IF) semi-circles. These semi-circles have been affected by changes in atmosphere. A suitable equivalent circuit has been used for extracting R and C values from the semi-circles; these values are tabulated in Table II. For all samples, $R_{HF}$ and $R_{IF}$ in H$_2$O-saturated atmosphere is the lowest; suggesting proton conductivity in all samples. For all samples, $C_{HF}$ in different atmospheres are of the order of $10^{-12}$ F/cm; suggesting HF semi-circle originates from the bulk of the sample[19]. On the other hand, $C_{IF}$ in different atmospheres are of the order of $10^{-11}$-$10^{-10}$ F/cm. This indicates the IF semi-circle originates predominately from minor phase and grain boundary [19].

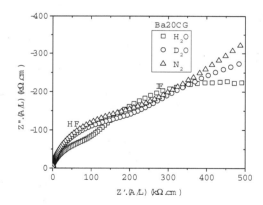

Figure 3. Impedance spectra of composite sample Ba20CG recorded at 200°C in different atmospheres.

Table II. Resistivity and capacitance values extracted from the HF and IF semi-circles.

| Sample | Atmosphere | High frequency | | Intermediate frequency | |
|---|---|---|---|---|---|
| | | Resistivity $R_{HF}$ (k$\Omega$.cm) | Capacitance $C_{HF}$ ($10^{-12}$ F/cm) | Resistivity $R_{IF}$ (k$\Omega$.cm) | Capacitance $C_{IF}$ ($10^{-10}$ F/cm) |
| Ba10CG | H$_2$O | 85 | 7.5 | 155 | 2.2 |
| | D$_2$O | 101 | 6.2 | 185 | 3.3 |
| | N$_2$ | 92 | 5.7 | 283 | 4.2 |
| Ba20CG | H$_2$O | 64 | 9.0 | 728 | 0.6 |
| | D$_2$O | 167 | 7.7 | 991 | 1.0 |
| | N$_2$ | 214 | 8.3 | 3036 | 2.9 |
| Ba30CG | H$_2$O | 46 | 6.5 | 683 | 1.0 |
| | D$_2$O | 88 | 6.6 | 1219 | 1.1 |
| | N$_2$ | 90 | 6.7 | 4304 | 1.6 |
| Ba40CG | H$_2$O | 30 | 5.7 | 442 | 1.7 |
| | D$_2$O | 64 | 5.8 | 964 | 1.7 |
| | N$_2$ | 55 | 5.6 | 4137 | 1.7 |

All samples show variation in bulk conductivity with atmosphere over the temperature range of measurement. The variation of conductivity in H$_2$O and D$_2$O-saturated atmospheres suggests protonic conduction in these samples over this temperature range. Temperature dependent conductivity of Ba20CG in different atmospheres is shown in Figure 4. The activation energy values extracted from the bulk conductivity Arrhenius plots are listed in Table III. The activation energy for the bulk conductivity of Ba10CG is significantly high and remains constant in different atmospheres. This indicates that the bulk charge transport in this sample is predominately controlled by the CG phase, which has significantly higher activation energy for the bulk conductivity than the BCG phase[5,6,14,15]. In contrast, the activation energy for the bulk conductivity of Ba20CG shows significant changes in different atmospheres. The activation energy in H$_2$O-saturated atmosphere is the lowest. In H$_2$O-saturated atmosphere, the protonic contribution to the bulk conductivity in the BCG phase (low activation energy) is significantly higher than the CG phase (high activation energy)[14,15,20]. Hence in Ba20CG, in H$_2$O-saturated atmosphere, the BCG phase controls the charge transport. The activation energy in dry nitrogen atmosphere is the highest; indicating that the charge transport is controlled by the BCG and CG phase. For samples Ba30CG and Ba40CG, difference between activation energy in D$_2$O and H$_2$O-saturated atmospheres is 0.04 eV. This is in agreement with calculated and observed values reported in the literature[14,15]. This confirms the bulk charge transport in these composites is predominately controlled by the protonic conductivity in the BCG phase in the low temperature range.

Figure 5 shows representative impedance spectra of composite samples at 360°C measured in different atmospheres. The impedance spectra show IF and low frequency (LF) semi-circles in different atmospheres and these semi-circles are displaced from the origin of Z' axis. The displacement along Z' axis is attributed to the bulk resistivity. Due to the composite nature of electrolytes; LF semi-circle is quiet likely to originate from the electrolyte. This has been confirmed by varying the A/L ratio of the sample. For samples Ba30CG and Ba40CG, the IF and LF semi-circles show significant variation in H$_2$O-saturated, D$_2$O-saturated and dry nitrogen atmospheres. This suggests both the IF and LF semi-circles have been affected by the conductivity variation in BCG grain boundary in different atmospheres. The resistivity and capacitance values extracted from the IF and LF semi-circles are listed in Table IV. For all samples, in H$_2$O atmosphere $C_{LF}$ is 2-3 orders of magnitude higher than the $C_{IF}$. This indicates the LF semi-circle originates from BCG-CG interface. Furthermore, $R_{LF}$ in H$_2$O-saturated atmosphere is the lowest; this can be a reason for high $C_{LF}$. On the other hand, in dry N$_2$ atmosphere $R_{LF}$ is the highest and $C_{LF}$ is the lowest. This is quiet likely due to increase in the resistivity of BCG grain boundary and its relaxation frequency shifting to lower values. The total resistivity is the intercept of LF semi-circle with the Z' axis.

The activation energy values extracted from the temperature dependent IF and LF conductivities are listed in Table V. Both Ba30CG and Ba40CG show similar type of variation in $E_{IF}$. On the other hand, the activation energy for the grain boundary conductivity of pure BCG phase do not vary significantly in different atmosphere[14,21]. Hence, IF conductivity is likely to be controlled CG phase and BCG grain boundary. In Ba30CG and Ba40CG $E_{LF}$ in H$_2$O-saturated atmosphere is lower than dry atmosphere. This indicates charge transport has been controlled by BCG-CG interface in H$_2$O-saturated atmosphere and BCG grain boundary in dry nitrogen atmosphere.

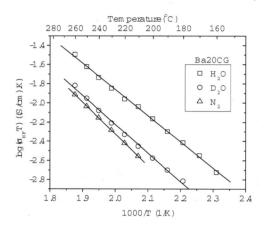

Figure 4. Temperature dependent HF conductivity plots of samples Ba20CG in different atmospheres.

Table III. The activation energy for the bulk conductivity in different atmospheres.

| Sample code | Activation energy (eV) | | |
|---|---|---|---|
| | H$_2$O | D$_2$O | N$_2$ |
| Ba10CG | 0.82 | 0.80 | 0.84 |
| Ba20CG | 0.54 | 0.60 | 0.66 |
| Ba30CG | 0.54 | 0.58 | 0.58 |
| Ba40CG | 0.50 | 0.54 | 0.54 |

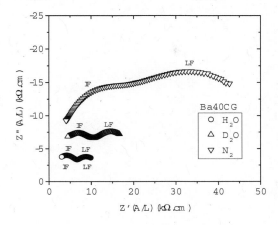

Figure 5. Impedance spectra of composite sample Ba40CG recorded at 360°C in different atmospheres. Each spectra is shifted along Z" axis to improve the clarity.

Table IV. Resistivity and capacitance values extracted from the IF and LF semi-circles.

| Sample | Atmosphere | Intermediate frequency | | Low frequency | |
|---|---|---|---|---|---|
| | | Resistivity $R_{IF}$ ($k\Omega.cm$) | Capacitance $C_{IF}$ ($10^{-10}$ F/cm) | Resistivity $R_{LF}$ ($k\Omega.cm$) | Capacitance $C_{LF}$ ($10^{-8}$ F/cm) |
| Ba10CG | $H_2O$ | 2.2 | 2.9 | 2.0 | 8.2 |
| | $D_2O$ | 2.2 | 4.7 | 1.6 | 14.7 |
| | $N_2$ | 3.2 | 8.6 | N/A | N/A |
| Ba20CG | $H_2O$ | 3.2 | 1.5 | 6.9 | 4.4 |
| | $D_2O$ | 3.6 | 2.9 | 6.9 | 4.9 |
| | $N_2$ | 7.9 | 7.7 | 3.6 | 5.3 |
| Ba30CG | $H_2O$ | 5.4 | 8.4 | 9.6 | 4.6 |
| | $D_2O$ | 7.0 | 1.2 | 13.8 | 4.4 |
| | $N_2$ | 19.9 | 5.7 | 18.7 | 2.4 |
| Ba40CG | $H_2O$ | 5.3 | .84 | 4.8 | 10.2 |
| | $D_2O$ | 8.5 | .91 | 8.8 | 6.6 |
| | $N_2$ | 17.3 | 3.5 | 31.2 | 1.5 |

Table V. Activation energy for the IF and LF conductivities in different atmospheres.

| Sample code | Atmosphere | Activation energy (eV) | |
|---|---|---|---|
| | | $E_{IF}$ | $E_{LF}$ |
| | | Intermediate Frequency | Low Frequency |
| Ba30CG | H$_2$O | 0.84 | 0.76 |
| | D$_2$O | 0.88 | 0.86 |
| | N$_2$ | 1.08 | 0.92 |
| Ba40CG | H$_2$O | 0.74 | 0.52 |
| | D$_2$O | 0.80 | 0.62 |
| | N$_2$ | 0.96 | 0.92 |

Sample Ba10CG does not show any noticeable variation in the total conductivity in different atmospheres over the temperature range of measurement. This is consistent with the high volume fraction of CG phase. Sample Ba20CG exhibits a small difference in the total conductivity at lower temperatures depending upon atmosphere, indicating presence of protonic conduction. However, above 500°C, the total conductivity measured in different atmospheres coalesces in to a single value indicating proton conductivity is not significant. Samples Ba30CG and Ba40CG show (Figure 6) considerable variation in the total conductivity in different atmospheres in the low temperature range; indicating significant contribution from the protonic conductivity. However, with the increasing temperature, the variation in resistance with atmosphere decreases indicating decreasing role of proton conduction. The activation energy for the total conductivity of samples in different atmospheres is listed in Table VI. For Ba10CG and Ba20CG E$_T$ do not show any significant change in different atmospheres. This shows the total conductivity for these composites is predominately controlled by the CG phase in this temperature of measurement in different atmospheres. For Ba30CG, E$_T$ in H$_2$O-saturated atmosphere is lower than dry N$_2$ atmosphere. This indicates BCG phase make contribution to the total conductivity in the temperature of measurement. The activation energy for the total conductivity of Ba40CG shows significantly low activation energy in wet atmosphere compared to dry atmospheres. This indicates in wet atmosphere total conduction has been controlled by BCG grain and BCG-CG interface. On the other hand, in dry atmosphere total conduction has been controlled BCG grain boundary and CG phase.

Table VI. The activation energy for the total conductivity in different atmospheres.

| | Activation energy $E_T$ (eV) | | |
|---|---|---|---|
| sample code | H$_2$O | D$_2$O | N$_2$ |
| Ba10CG | 0.84 | 0.84 | 0.84 |
| Ba20CG | 0.86 | 0.86 | 0.90 |
| Ba30CG | 0.78 | 0.82 | 0.94 |
| Ba40CG | 0.62 | 0.68 | 0.88 |

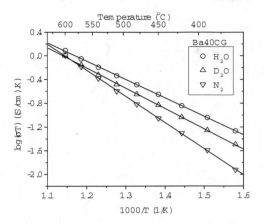

Figure 6. Temperature dependent total conductivity plots of sample Ba40CG in different atmospheres.

CONCLUSIONS

For composites with $x \geq 0.2$ show the highest conductivity in H$_2$O-saturated atmosphere in the temperature range 160°C-260°C. Furthermore, the difference between bulk activation energy in D$_2$O-saturated atmosphere and H$_2$O saturated atmosphere confirm bulk proton conduction in the temperature range 160°C-260°C. In the temperature range 350°C-600°C, the total conductivity of composites with x=0.4 has been controlled predominately by the bulk BCG phase and BCG-CG interface in H$_2$O saturated atmosphere. On the other hand, in dry nitrogen atmosphere the total conductivity has been controlled by CG and BCG grain boundary.

REFERENCES

[1]O. Yamamoto, Solid oxide fuel cells: fundamental aspects and prospects, *Electrochemica Acta*, **45,** 2423-35 (2000).

[2]A. B. Stambouli and E. Traversa, Solid oxide fuel cells (SOFCs): a review of an environmentally clean and efficient source of energy, *Renewable and Sustainable Energy Reviews,* **6,** 6433-55 (2002).

[3]N. Q. Minh, Ceramic fuel cells, *J. Am. Ceram. Soc.*, **76,** 563-88 (1993).

[4]Y. Arachi, H. Sakai, O. Yamamoto, Y. Takeda, and N. Imanishai, Electrical conductivity of the ZrO$_2$-Ln$_2$O$_3$ (Ln=lanthanides) system, *Solid State Ionics*, **121,** 133-9 (1999).

[5]K. Eguchi, T. Setoguchi, T. Inoue, and H. Arai, Electrical properties of ceria-based oxides and their application to solid oxide fuel cells, *Solid State Ionics,* **52,** 165-72 (1992).

[6]R. S. Torrens, N. M. Sammes, and G. A. Tompsett, Characterisation of (CeO$_2$)$_{0.8}$(GdO$_{1.5}$)$_{0.2}$ synthesised using various techniques, *Solid State Ionics,* **111,** 9-15 (1998).

[7] T. Kudo and H. Obayashi, Mixed electrical conduction in the Flourite-type $Ce_{1-x}Gd_xO_{2-x/2}$, *J. Electorochem. Soc.*, **123**, 415-9 (1976).

[8] N. Taniguchi, K. Hatoh, J. Niikura, T. Gamo, and H. Iwahara, Proton conductive properties of gadolinium-doped barium cerates at high temperatures, *Solid State Ionics*, **53-56**, 998-1003 (1992).

[9] N. Bonanos, Transport properties and conduction mechanism in high-temperature protonic conductors, *Solid State Ionics*, **53-56**, 967-74 (1992).

[10] H. Iwahara, T. Yajima, T. Hibino, and H. Ushida, Performance of solid oxide fuel cell using proton and oxide ion mixed conductors based on $BaCe_{1-x}Sm_xO_{3-\alpha}$, *J. Electrochem. Soc.*, **140**, 1687-91 (1993).

[11] T. Hibino, A.Hashimoto, M. Suzuki, and M. Sano, A solid oxide fuel cell using Y-doped $BaCeO_3$ with Pd-Loaded FeO anode and $Ba_{0.5}Pr_{0.5}CoO_3$ cathode at low temperatures, *J. Electrochem. Soc.*, **149**, A1503-8 (2002).

[12] M. S. Islam, Ionic transport in $ABO_3$ perovskite oxides: a computer modeling tour, *J. Mater. Chem.*, **10**, 1027-38 (2000).

[13] A. S. Nowick and Y. Du, *Solid State Ionics* **77**, 137 (1995).

[14] S. M. Haile, D. L. West, and J. Campbell, The role of microstructure and processing on the proton conducting properties of gadolinium-doped barium cerate, *.J. Mater. Res.*, **13**, 1576-95 (1998).

[15] D. A. Stevenson, N. Jiang, R. M. Buchanan, and F. E. G. Henn, Characterization of Gd, Yb and Nd doped barium cerates as proton conductors, *Solid State Ionics*, **62**, 279-85 (1993).

[16] A. Venkatasubramanian, P. Gopalan, and T. R. S. Prasanna, Synthesis and characterization of electrolytes based on $BaO-CeO_2-GdO_{1.5}$ system for intermediate temperature solid oxide fuel cells, *Int. J. Hydrogen Energy*, **35**, 4597-605 (2010).

[17] www.norecs.com

[18] D. Johnson, ZView software, 3.0 Version, Scribner Associates.

[19] J. G. Fletcher, A. R. West, and J. T. S. Irvine, The AC impedance response of the physical interface between yttria-stabilized zirconia and $YBa_2Cu_3O_{7-x}$. *J. Electrochem. Soc.*, **142**, 2650-54 (1995).

[20] B. Zhu, I. Albinsson, and B.E. Mellander, Electrical Properties and Proton Conduction of Gadolinium Doped Ceria. *Ionics*, **4**, 261-266 (1998).

[21] J. Wu, PhD thesis, California Institute of Technology, 86 (2005).

# 3D CFD ANALYSIS FOR SOLID OXIDE FUEL CELLS WITH FUNCTIONALLY GRADED ELECTRODES

Junxiang Shi and Xingjian Xue[*]
Department of Mechanical Engineering
University of South Carolina, Columbia, SC 29208, USA

## ABSTRACT

A comprehensive 3D CFD model is developed for a SOFC and includes complicated transport phenomena of mass/heat transfer, charge (electron and ion) migration, and electrochemical reactions. The uniqueness of this study is that functionally graded porous electrode property is taken into account, which involves not only linear but also nonlinear electrode porosity distributions in a general sense. The numerical code is validated using experimental data from open literature. Extensive numerical analysis is performed to elucidate both porous microstructure distributions and operating condition effects on cell performance from theoretical point of view. Results indicate that the cell with inverse parabolic porous electrode demonstrates promising performance; the optimal profile of inverse parabolic porosity distribution is dependent on operating conditions, typically pressure losses across electrodes. Increasing electrode thickness makes the V-I curve approach toward each other for the cells with linear and inverse parabolic porous electrodes. This trend is not apparent for the cell with parabolic porous electrode. In general, increasing electrolyte thickness leads to worse cell performance. Fuel/gas flow settings influence both cell temperature distribution and cell performance.

## 1. INTRODUCTION

Solid oxide fuel cell (SOFC) has been widely recognized as one of promising clean energy conversion devices that convert the chemical energy of hydrogen into electrical energy directly. The fundamental structure of SOFCs is positive electrode-electrolyte-negative electrode (PEN) assembly, where dense electrolyte is sandwiched by porous electrode on either side. Based on this fundamental PEN assembly, two classical SOFC designs, planar and tubular, have been extensively investigated both experimentally and numerically in literature [1-6]. While experimental approach plays a significant role in SOFC technique advancement, it is generally very costly and time-consuming. Furthermore, it is very difficult to directly measure the detailed profile of multi-physics processes in a cell/stack. Modeling and numerical analysis has emerged as a cost-effective approach to de-convolute the complicated transport phenomena in SOFCs and facilitate prototype development and pre-experimental analysis.

In literature, mathematical SOFC models have been developed at different levels [7-22], including 1-D models [7], 2-D models [8-10], and 3-D models [11-16], among which 3-D computational fluid dynamics (CFD) modeling approach plays an increasingly significant role for the analysis and design of practical SOFC systems. Ahmed [11] developed a mathematical model to simulate electrochemical and thermal behavior of a SOFC stack with honeycomb structure. Recknagle et al. [12] presented a 3D model to simulate a SOFC with three flow settings, i.e., co-flow, counter-flow and cross-flow designs. Results shown that the co-flow case led to the most uniform temperature distributions and the smallest thermal gradients, and consequently offered thermal-structural advantages over the other flow cases. Inui et al. [13] proposed a new cell temperature control method to optimize the operating parameters by minimizing the cell temperature shift from its normal value. Larrain et al. [17] investigated temperature field and active area of a SOFC through a combined

---

[*] Corresponding author: Tel.: 1-803-576-5598; Fax: 1-803-777-0106; Email Address: Xue@cec.sc.edu (X. Xue)

thermal and kinetic model. It is worth mentioning that porosity distribution of electrodes is generally assumed uniform in these investigations.

Recently Cable and Sophie [18] created a novel design of symmetrical, bi-electrode supported cell (BSC). The uniqueness of this design is functionally graded porous electrodes fabricated using freeze tape casting technique. While experimental results shown that BSC with functionally graded porous electrodes may significantly improve cell performance and power density [18], the fundamental mechanism is not well understood. The heterogeneous porosity feature of electrode renders the conventional SOFC modeling method inadequate, where electrode properties are generally assumed homogeneous. Greene et al. [19] investigated an electrode with a functionally graded porosity distribution in 1-D setting with the mean-transport pore model to approximate the geometry of a porous media. Results shown that increasing electrode porosity near the electrolyte improved the cell performance for certain fuels. Ni et al. [20] studied particle size grading and consequently porosity grading effects on SOFC performance using 1-D model, in which the percolation theory is employed for fundamental parameter characterization. The results demonstrated that while this design could significantly enhance the gas transport for thick electrodes, too much grading for thin electrodes may increase the activation over-potentials. It is also claimed that particle size grading is generally more effective than porosity grading. It is worth noting that these two investigations just assume linear porosity gradient in 1-D settings. Clearly the modeling results from these two studies are not applicable for BSC design, where fuel/gas flow is designed to be perpendicular to porosity gradient; what's more, the graded porosity in this design is not necessarily linear. This creates a need to develop a SOFC model that can handle heterogeneous porous electrode properties in a general sense.

Inspired by BSC designs, in this research, we investigate heterogeneous electrode property effect on SOFC performance in a general sense through 3D comprehensive CFD modeling. In particular, a variety of functionally graded electrode porosity distributions and their influence on SOFC performance are studied theoretically. The model includes complicated interactions among charge (electron/ion) transport, mass/heat transfer, and electrochemical reactions. The model developed in this paper can be utilized to optimize porous electrode design.

## 2. MODEL DESCRIPTION

The schematic of tri-layer cell based stack is shown in Figure 1. The cell is composed of thin layer electrolyte sandwiched by symmetric electrode on either side, where the electrode is functionally graded porous microstructure. In functionally graded electrode development, ceramic powder particles are mixed with solvent (water). After freezing, subliming, and sintering, porous ceramic particle packing is obtained with multi-scale porosity percolated by ceramic particle backbone. Given the powder particles being at the order of micro-meter or even smaller, porous ceramic backbone obtained through freeze tape casting is essentially a random packing of powder particles with principal porosity being functionally graded in the direction perpendicular to the electrolyte layer. Based on this consideration, it is assumed that the percolation theory is still applicable for fundamental parameter characterization for functionally graded electrode. The porous electrode can be infiltrated with active electrode materials to form functional

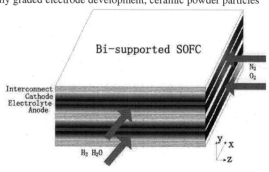

Fig. 1 Schematic of BSC stack unit

composite electrode; alternatively, composite electrode can be formed by mixing electronic conductor particles and ionic conductor particles during freeze tape casting process. Without loss of generality, we consider composite electrodes composed of electronic conductor particles, ionic conductor particles, and void phase. We assume that the fundamental physical properties of porous electrode such as thermal conductivity, permeability, etc., can be described using widely utilized empirical formulations such as in [23]. In the model development, it is also assumed that humidified hydrogen is used as a fuel, while air is used for cathode gas. A typical mathematical SOFC model, examining electrochemical, charge, flow, and thermal effects, considers three different domains: cathode electrode Layer, electrolyte Layer and anode electrode layer, and includes the coupled processes of charge (ion/electron) balance, electrochemical reaction kinetics, mass balance, and momentum and energy balances. The model development is detailed as follows.

2.1 Charge Balance

Charge transport includes ionic and electronic transports. With the assumption of composite electrode, both electronic and ionic transports are allowed in anode and cathode electrodes, while electrolyte only allows ions to migrate through. According to the Ohm's law, the governing equations for charge balance can be described as [7, 21]:

$$\text{Electronic charge:} \begin{cases} \text{Anode electrode layer} : \nabla \cdot \left( \sigma_a^{eff} \nabla \varnothing_e \right) = -j_a A_V \\ \text{Cathode electrode layer} : \nabla \cdot \left( \sigma_c^{eff} \nabla \varnothing_e \right) = -j_c A_V \end{cases} \tag{1}$$

$$\text{Ionic charge:} \begin{cases} \text{Electrolyte layer} : \nabla \cdot \left( \kappa \nabla \varnothing_i \right) = 0 \\ \text{Anode electrode layer} : \nabla \cdot \left( \kappa_a^{eff} \nabla \varnothing_i \right) = j_a A_V \\ \text{Cathode electrode layer} : \nabla \cdot \left( \kappa_c^{eff} \nabla \varnothing_i \right) = j_c A_V \end{cases} \tag{2}$$

Where the effective ionic and electronic conductivities in the electrode layers are defined as [9, 10, 22]:

$$\kappa^{eff} = \kappa[(1-\phi)(1-\varepsilon)P_{io}]^m \tag{3}$$

$$\sigma^{eff} = \kappa[\phi(1-\varepsilon)P_{el}]^m \tag{4}$$

Here $P_i$ is the probability for $i$-phase particles to form percolated or globally continuous clusters and can be calculated as [24]:

$$P_i = [1 - \left( \frac{4.236 - Z_{i-i}}{2.472} \right)^{2.5} ]^{0.4} \tag{5}$$

The exponent $m$ in equations (3) and (4) depends on the distribution of the $i$-phase material in space [9].

The coordination number $Z_{i-j}$ between $i$-phase particles and $j$-phase particles in Equation (5) can be expressed as [7, 9, 10]:

$$Z_{i-j} = n_j \frac{Z_i Z_j}{Z} \tag{6}$$

Here Z is the total average coordination number, which is equal to 6 in general [7, 9, 10]; $Z_i$ is the total coordination for $i$-phase particles [25]. And the total coordination for electronic conducting phase and ionic conducting phase can be represented using Equation (7) and (8) respectively.

$$Z_{el} = 3 + \frac{Z - 3}{n_{el} - (1 - n_{el})(\frac{d_{io}}{d_{el}})^2} \tag{7}$$

$$Z_{io} = 3 + \frac{Z-3}{n_{el}(\frac{d_{el}}{d_{io}})^2 - (1-n_{el})} \tag{8}$$

The number fraction $n_i$ is related to volume fraction of $i$-phase, and can be represented using Equation (9) and (10) for electronic conducting phase and ionic conducting phase respectively.

$$n_{el} = \frac{(\frac{d_{io}}{d_{el}})^3 \varnothing_{el}}{1 - \varnothing_{el} + (\frac{d_{io}}{d_{el}})^3 \varnothing_{el}} \tag{9}$$

$$n_{io} = 1 - n_{el} \tag{10}$$

The volumetric reactive surface area $A_V (m^{-1})$ in Equations (1) and (2) is determined by the number of percolated electronic particles contained in the thickness $d_{el}$ as [20, 26]:

$$A_V = \pi \sin^2 \sin^2(\theta_c) N_t d_{el} d_{io} n_{el} n_{io} P_{io} P_{el} \frac{Z_{el} Z_{io}}{Z} \tag{11}$$

Here $N_t$ is the number density of all particles, given as

$$N_t = \frac{1-\varepsilon}{\left(\frac{4\pi}{3}\right) d_{el}^3 (n_{el} + (1-n_{el})\left(\frac{d_{io}}{d_{el}}\right)^3)} \tag{12}$$

$j_a$ and $j_c$ in Equations (1) and (2) are the volumetric current densities generated in the anode and cathode layers due to the $H_2$ oxidation and $O_2$ reduction reactions respectively, which can be calculated using the general Butler-Volmer equation:

$$j_a = j_{0,ref}^{H_2} \left(\frac{c_{H_2}}{c_{H_2,ref}}\right)^{\gamma_{H_2}} [\exp\exp\left(\frac{\alpha n F \eta_a}{RT}\right) - \exp\exp\left(-\frac{(1-\alpha)nF\eta_a}{RT}\right)] \tag{13}$$

$$j_c = j_{0,ref}^{O_2} \left(\frac{c_{O_2}}{c_{O_2,ref}}\right)^{\gamma_{O_2}} [\exp\exp\left(\frac{\alpha n F \eta_c}{RT}\right) - \exp\exp\left(-\frac{(1-\alpha)nF\eta_c}{RT}\right)] \tag{14}$$

Here, $\eta_i$ is the activation overvoltage and can be expressed as:

$$\eta_i = \varnothing_e - \varnothing_i - \Delta\varnothing_{eq} \tag{15}$$

Here $\Delta\varnothing_{eq}$ is the equilibrium potential difference (V).

2.2 Mass Conversation

The multi-species transport phenomena can be described using modified Stefan-Maxwell equations. And the Knudsen diffusion [7, 9, 10] is employed here to take into account the porous effect in porous electrode layers:

$$\nabla \cdot \left( \rho u \omega_i - \rho \omega_j \sum D_{ij}^{eff} (\nabla x_i + (x_i - \omega_j)\frac{\nabla p}{p}) \right) = S_i \tag{16}$$

Where $S_i$ is the reaction source term for species $i$, $\omega_i$ the weight fraction of species $i$, $x_i$ the mole fraction of species $i$. $D_{ij}^{eff}$ is the effective diffusion coefficient and is written as:

$$D_{ij}^{eff} = \frac{\varepsilon}{\tau}(\frac{D_{ij} D_{Kn,i}}{D_{ij} + D_{Kn,i}}) \tag{17}$$

Here $\varepsilon$ is electrode porosity, $\tau$ the tortousity, $D_{ij}$ is the binary diffusivity coefficient for a pair of species $i$ and $j$, and can be calculated as:

$$D_{ij} = \frac{1.43e^{-8}T^{1.75}}{pM_{ij}^{\frac{1}{2}}(V_i^{\frac{1}{3}}+V_j^{\frac{1}{3}})} \tag{18}$$

Here $M_{ij}$ is the mean molecular mass

$$M_{ij} = \frac{2}{\dfrac{1}{M_i}+\dfrac{1}{M_j}} \tag{19}$$

$D_{Kn}$ is the Knudsen diffusion coefficient of species $i$.

$$D_{Kn,i} = \frac{97}{2}d_{pore}\sqrt{\frac{T}{M_i}}, d_{pore} = \frac{2}{3}\frac{\varepsilon}{1-\varepsilon}d_p \tag{20}$$

The average molecular weight is calculated as:

$$M = \sum_{j=1}^{n}x_jM_j \tag{21}$$

When ideal gas is considered, the density can be written as:

$$\rho = \frac{pM}{RT} \tag{22}$$

2.3 Momentum Equations

Due to high temperature operating conditions, the weakly compressible Brinkman equations are assumed to govern fluid flow in porous electrodes.

$$\left(\frac{\mu}{K}+S\right)u = \nabla\cdot\left[-pI+\frac{\mu}{\varepsilon}\left((\nabla u+(\nabla u)^{T})-\frac{2}{3}(\nabla\cdot u)I\right)\right] \tag{23}$$

And continuity equation in porous electrodes:

$$\nabla\cdot(\rho u) = S \tag{24}$$

Here $I$ is the unit matrix, $S$ is the mass source term determined by current density in reaction zone.

$$S = \sum_i\frac{j_iM_i}{n_iF} \tag{25}$$

And $K$ is the flow permeability determined using Carman-Kozeny correlation as:

$$K = \frac{\varepsilon^3 d_p^3}{180(1-\varepsilon)^2} \tag{26}$$

2.4 Energy Conservation

For energy conservation, thermal equilibrium model is employed in this research, where the temperature of fluid is the same as that of porous solid in the volumetric average sense. The corresponding equation can be expressed as [22]:

$$\nabla\cdot(k\nabla T+\sum_i h_in_i)+\rho C_pu\cdot\nabla T = S_h^e+S_h^j \tag{27}$$

Where $S_h^e$ and $S_h^j$ represent heat source terms that account for the heat generated by the electrochemical reaction and joule heating, and can be written as:

$$S_h^e = \left(\Delta E_{elec-chem}-\Delta G\right)\frac{j}{n_iF}, S_h^j = \sigma_{io}\varphi_{io}^2 \tag{28}$$

## 2.6 Boundary Conditions

The boundary conditions are needed to solve the coupled partial differential equations mentioned above, including inlet/outlet boundary conditions, continuity of flux, impermeability and no slip assumptions, etc., which are summarized in Table 1 for co-flow settings. For other flow settings such as cross-flow and counter-flow, the corresponding boundary conditions can be similarly obtained.

Table 1 Boundary condition settings in numerical computation (Co-flow)

| Location | Boundary Conditions (Charge, Mass, Momentum, Energy) |
|---|---|
| Cathode inlet | $\partial \phi_e / \partial z = 0$, $x_j = x_{j,ca,inlet}$, $p = p_{out} + \Delta p_c$, $T = T_{op}$ |
| Anode inlet | $\partial \phi_e / \partial x = 0$, $x_j = x_{j,an,inlet}$, $p = p_{out} + \Delta p_a$, $T = T_{op}$ |
| Cathode outlet | $\partial \phi_e / \partial z = 0$, $p = p_{out}$, $\partial x_j / \partial z = 0$, $\partial T / \partial z = 0$ |
| Anode outlet | $\partial \phi_e / \partial x = 0$, $p = p_{out}$, $\partial x_j / \partial x = 0$, $\partial T / \partial x = 0$ |
| Cathode/Interconnect | $\phi_e = V_{cell}$, $\partial x_j / \partial y = 0$, $u = 0$, $\partial T / \partial y = 0$ |
| Anode /Interconnect | $\phi_e = 0$, $\partial x_j / \partial y = 0$, $u = 0$, $\partial T / \partial y = 0$ |
| Cathode/Electrolyte | $\partial \phi_e / \partial y = 0$, $\partial \phi_i / \partial y = 0$, $\partial x_j / \partial y = 0$, $u = 0$, $\partial T / \partial y = 0$ |
| Anode / Electrolyte | $\partial \phi_e / \partial y = 0$, $\partial \phi_i / \partial y = 0$, $\partial x_j / \partial y = 0$, $u = 0$, $\partial T / \partial y = 0$ |

## 3. MODEL VALIDATION

The mathematical model presented in this paper is numerically solved using finite element package COMSOL MULTIPHYSICS version 3.5. Due to the lack of BSC experimental data in open literature, the model code is validated using other SOFC experimental data [27]. The purpose of this validation is to examine our numerical code. The physical parameters used in the model validation are shown in Table 2. The comparison results are shown in Figure 2. It can be seen that the numerical results agree very well with experimental results.

Table 2 Parameters used for SOFC model validation [7, 9]

| | |
|---|---|
| Surrounding temperature, $T_{op}$ (K) | 1073.0 |
| Outlet pressure of electrodes, $p_{out}$ (atm) | 1.0 |
| Fuel composition, $x_{H_2} : x_{H_2O}$ | 0.95:0.05 |
| Air composition, $x_{O_2} : x_{N_2}$ | 0.21:0.79 |
| Anode conductivity, $\sigma$ (S/m) | 71428.57 |
| Cathode conductivity, $\sigma$ (S/m) | 5376.34 |
| Electrolyte conductivity, $k$ (S/m) | 0.64 |
| Cell length (cm) | 0.1 |
| Cell Width (cm) | 0.1 |
| Anode electrode layer thickness $t_a$ ($\mu$m) | 1000.0 |
| Cathode electrode layer thickness $t_c$ ($\mu$m) | 70 |
| Electrolyte thickness $t_{elec}$ ($\mu$m) | 8 |
| Anode Channel $t_{a,ch}$ ($\mu$m) | 1000.0 |
| Cathode Channel $t_{c,ch}$ ($\mu$m) | 1000.0 |
| Porosity of anode and cathode $\varepsilon$ | 0.25 |
| Tortuosity of anode and cathode $\tau$ | 1.5 |
| Particle diameter $d_p$ ($\mu$m) | 0.5 |

| | |
|---|---|
| Contact angel between e⁻ and O²⁻ conducting particles $\theta(°)$ | 30 |
| Radius of e⁻ conducting particles $r_{el}$ (µm) | 0.1 |
| Radius of O²⁻ conducting particles $r_{io}$ (µm) | 0.1 |
| Volume fraction of e⁻ conducting particles $\phi$ | 0.5 |
| Reference $H_2$ concentration $c_{H_2,ref}$ | 10.78 |
| Reference $O_2$ concentration $c_{O_2,ref}$ | 2.38 |
| Reaction order for $H_2$ oxidation $\gamma_{H_2}$ | 0.5 |
| Reaction order for $O_2$ reduction $\gamma_{O_2}$ | 0.5 |

## 4. RESULTS AND DISCUSSION

The validated numerical code is then utilized to investigate SOFC performance with functionally graded electrode in a general sense. The physical parameters used in the simulations are listed in Table 3. Here a relatively small cell is utilized to illustrate simulation and analysis in order to reduce computational cost, the identical procedure can be utilized for large cell analysis. The porosity distribution of electrodes in the simulation is approximated with a function of variable y as listed in Table 3 to take into account the feature of functionally graded porosity. The SOFC performance variations induced by a wide range of pressure losses through porous electrode are also investigated from pure theoretical point of view.

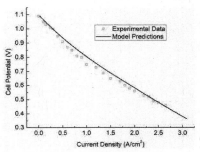

Fig. 2 Comparison between numerical results and experimental data

Table 3 Parameters used in the simulation [7, 9]

| | |
|---|---|
| Surrounding temperature, $T_{op}$ (K) | 1073.0 |
| Outlet pressure of electrodes, $p_{out}$ (atm) | 1.0 |
| Fuel composition, $x_{H_2} : x_{H_2O}$ | 0.95:0.05 |
| Air composition, $x_{O_2} : x_{N_2}$ | 0.21:0.79 |
| Anode conductivity, $\sigma$ (S/m) | $(9.5 \times \frac{10^7}{T})e^{-\frac{1150}{T}}$ |
| Cathode conductivity, $\sigma$ (S/m) | $(4.2 \times \frac{10^7}{T})e^{-\frac{1200}{T}}$ |
| Electrolyte conductivity, $k$ (S/m) | $(3.34 \times \frac{10^4}{T})e^{-\frac{10300}{T}}$ |
| Interconnect conductivity, $k$ (S/m) | 35000 |
| Cell length (cm) | 0.5 |
| Cell width (cm) | 0.5 |
| Anode electrode layer thickness $t_a$ (µm) | 300 |
| Cathode electrode layer thickness $t_c$ (µm) | 300 |

| Electrolyte thickness $t_{elec}(\mu m)$ | 100 |
|---|---|
| Interconnect thickness $t_{inter}(\mu m)$ | 100 |
| Pressure loss on Anode layer $\Delta p_a$ | 100 |
| Pressure loss on Cathode layer $\Delta p_c$ | 750, 1500, 3000 |
| Porosity $\varepsilon$ | $f_\varepsilon(y)$ |
| Tortuosity $\tau$ | $\varepsilon^{-0.5}$ |
| Contact angel between $e^-$ and $O^{2-}$ conducting particles $\theta(^\circ)$ | 30 |
| Radius of $e^-$ conducting particles $r_{el}(\mu m)$ | 0.1 |
| Radius of $O^{2-}$ conducting particles $r_{io}(\mu m)$ | 0.1 |
| Volume fraction of $e^-$ conducting particles $\phi$ | 0.5 |
| Reference $H_2$ concentration $c_{H_2,ref}$ | 10.78 |
| Reaction order for $H_2$ oxidation $\gamma_{H_2}$ | 0.5 |
| Reaction order for $O_2$ reduction $\gamma_{O_2}$ | 0.5 |
| Anode thermal conductivity, k (W/(m K)) | 11 |
| Cathode thermal conductivity, k (W/(m K)) | 11 |
| Electrolyte thermal conductivity, k (W/(m K)) | 2.7 |
| Interconnect thermal conductivity, k (W/(m K)) | 2.5 |

4.1 Functionally-graded porosity distribution effect on SOFC performance

The uniqueness of BSC design is functionally graded porous electrode, where the porosity has the maximum value at the electrode/interconnect interface while reaching the minimum value at the electrode/electrolyte interface. Within electrodes, the porosity changes gradually (not necessarily linear) from the two extremes. According to this physical feature, three porosity distributions are assumed in a general sense, i.e., parabolic, linear, inverse parabolic, as expressed in Equation (29).

$$\begin{cases} f_1(y) = a_1 y^2 + c_1 \\ f_2(y) = a_2 y + c_2 \quad f_i^\varepsilon \in [0.1, 0, 9], i = 1, 2, 3 \\ f_3(y) = a_3 \sqrt{y} + c_3 \end{cases} \tag{29}$$

Where $a_i$ and $c_i$ are constants and determined according to the different electrode regions.

The porosity distributions shown in Equation (29) are then employed to investigate cell performance. Figure 3 shows cell performances (V-I curve) under the cathode electrode pressure loss conditions of 750Pa, 1500Pa and 3000Pa respectively. As one can see from Figure 3 (a) that the cell performances ordered from low to high in terms of electrode porosity distributions are parabolic, linear, and inverse parabolic, where 750Pa pressure loss is applied through the cathode. When pressure loss increases from 750Pa to 1500Pa, the performance of the cell with linear porous electrodes is significantly improved; it has little effect on the performance of the cells with parabolic and inverse parabolic porous electrodes, as shown in Figure 3 (b). When pressure loss is further increased from 1500Pa to 3000Pa, the change of cell performances shown in Figure 3 (c) has the similar trend to that in Figure 3 (b).

The corresponding mass concentration distributions shown in Figures 4-6 are employed to understand the fundamental mechanisms associated with porosity distribution effects on cell performance, where the cell voltage is set to 0.4V. Essentially the resulting oxygen concentration distributions are the combinational effects of both flow resistance through porous electrode and electrochemical reaction rate (oxygen consumption rate). According to electrode porosity distributions in Equation (29), one can estimate that the average flow resistance ordered from high to low in terms of porosity distributions are parabolic, linear, inverse parabolic. It is quite possible that the oxygen concentration gradients should be in the same order. As one can see from Figure 4 (a) and (b) that oxygen concentration distributions are quite similar to each other for parabolic and linear porosity distributions. One possible reason is that oxygen concentration distributions induced by flow resistance are balanced by the oxygen consumed through the electrochemical reactions. As a result, the oxygen consumption (thus current density) in linear porous electrode is more than that in parabolic porous electrode. This leads to that the current density of the cell with linear porous electrode is higher than that with parabolic porous electrode as shown in Figure 3 (a). The oxygen concentration distribution with inverse parabolic porous electrodes in Figure 4 (c) tends to be more uniform. It is not straightforward to compare with the other two cases. When comparing hydrogen distributions in Figure 4, one can see that the hydrogen distribution in the cell with both parabolic and linear porous electrodes (Figure 4 (a) and (b)) is more uniform than that with inverse parabolic porous electrode (Figure 4(c)). According to above analysis, one can infer that current density in Figure 4 (c) is higher than those in Figure 4 (a) and (b). This observation is actually consistent with those in Figure 3 (a). When pressure loss increases from 750Pa to 1500Pa and 3000Pa, the corresponding mass distributions are shown in Figure 5 and 6 respectively. And current density is dependent on mass distributions in a complex way and is not straightforward to quantify.

According to previous analysis results, one can see that the cell with inverse parabolic porosity distribution demonstrates very promising performance. To further elucidate porosity distribution effects on cell performance and optimize cell performance, the porosity distribution is generalized as follows:

$$f(y) = ay^N + c, \ N \in [0.15, 2] \tag{30}$$

Where $a$ and $c$ are constants and determined by the different assumptions of electrode porosity gradients. $N$ is a variable controlling the profile of porosity distribution. We then employ this generalized electrode porosity distribution to investigate cell performance with different values of $N$. Here the cell voltage is still set at 0.4V (cell performance can be similarly obtained under other voltage conditions), while the cell current density is calculated under three pressure loss conditions, i.e., 750Pa, 1500Pa, and 3000Pa. The results are shown in Figure 7. It can be seen that maximum current density is obtained when pressure loss is 3000Pa and the corresponding $N$ value is 0.35. With the decrease of pressure loss variations from 3000Pa to 1500Pa and 750Pa, the corresponding optimal N values are 0.25 and 0.2 respectively. Obviously optimal porosity distribution is operating condition dependent.

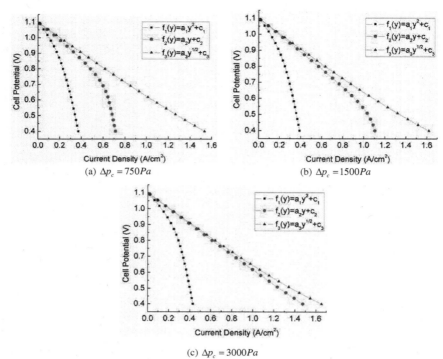

(a) $\Delta p_c = 750 Pa$      (b) $\Delta p_c = 1500 Pa$

(c) $\Delta p_c = 3000 Pa$

Fig. 3 Cell performance under different cathode pressure loss

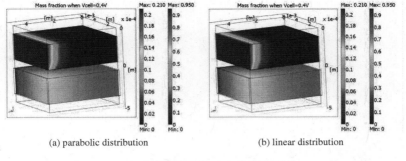

(a) parabolic distribution  (b) linear distribution

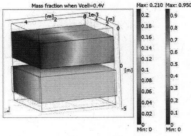

(c) inverse parabolic distribution

Fig. 4 Mass concentration under $\Delta p_c = 750 Pa$

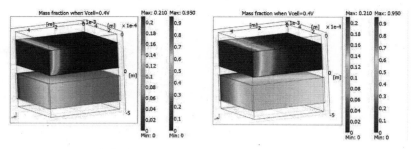

(a) parabolic distribution          (b) linear distribution

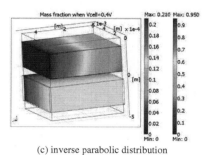

(c) inverse parabolic distribution

Fig. 5 Mass concentration under $\Delta p_c = 1500Pa$

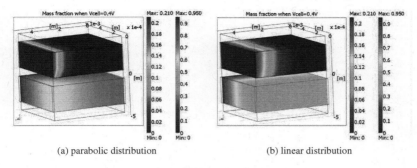

(a) parabolic distribution           (b) linear distribution

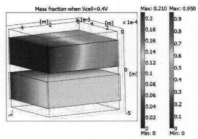

(c) inverse parabolic distribution

Fig. 6 Mass concentration under $\Delta p_c = 3000 Pa$

## 4.2 Effects of electrode and electrolyte thicknesses

Electrolyte layer is a critical component of SOFC. While thin electrolyte layer may lead to high (ionic) conductivity, it also imposes great challenges on fabrication process; thick electrolyte layer may potentially facilitate cell fabrication, it will induce high ionic resistance and reduced cell performance. Clearly the thickness of electrolyte needs a suitable tradeoff for practical SOFC development. On the other hand, electrode thickness may potentially influence the porosity distributions and thus cell performance. In this section, numerical analysis is performed to investigate electrode and electrolyte thickness effects on cell performance.

When electrode thickness increases

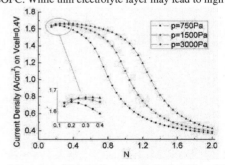

Fig. 7 Variation of Current Density with N when $V_{cell} = 0.4V$

from 300μm (the corresponding cell performance is shown in Figure 3) to 400μm, the rest of the conditions keep the same, the cell performance is shown in Figure 8. Comparing to the cell performance in Figure 3, one can see that increase electrode thickness has little effect on the performance of the cells with parabolic and inverse parabolic porosity distributions. The performance of the cell with linear porosity distribution is significantly improved under pressure loss conditions of 750Pa and 1500Pa as shown in Figure 3 (a) and (b), and Figure 8 (a) and (b). One also can see that the corresponding V-I curve is pushed toward that with inverse parabolic porosity distribution. The reason is that, when the thickness of electrode increases, the inverse parabolic porosity distribution is squeezed toward the linear distribution. As a result, the cell performance approaches toward each other, given the rest of the conditions are the same. Similarly increase electrode thickness also squeezes parabolic porosity distribution toward the linear distribution from the other side, the change of V-I curve is not significant, albeit, toward that with linear porosity distribution as shown in Figure 8.

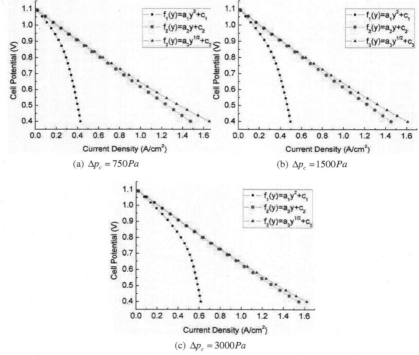

(a) $\Delta p_c = 750Pa$       (b) $\Delta p_c = 1500Pa$

(c) $\Delta p_c = 3000Pa$

Fig. 8 Cell performance with $t_a = t_c = 400\mu m$, $t_e = 100\mu m$

When the electrolyte thickness increases from 100μm (the corresponding cell performance is shown in Figure 3) to 200μm, the corresponding cell performance is shown in Figure 9. As one can see that the cell performance significantly decreases. In this situation, thicker electrolyte leads to ohmic loss so significant that increase pressure loss has little effect on cell performance.

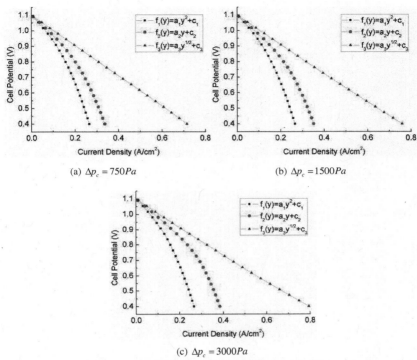

(a) $\Delta p_c = 750 Pa$

(b) $\Delta p_c = 1500 Pa$

(c) $\Delta p_c = 3000 Pa$

Fig. 9 Cell performance with $t_a = t_c = 300\mu m$, $t_e = 200\mu m$

## 4.3 Effects of thermal field

The thermal field effect on cell performance is also investigated, where inverse parabolic porosity distribution is employed, and the performance of the cell in isothermal conditions is compared with that in non-isothermal conditions. As one can see from Figure 10 that the same V-I curve is obtained for both isothermal and non-isothermal assumptions when the cell is operated in low current density conditions (less than $0.6A/cm^2$ in this case). Beyond the current density of $0.6A/cm^2$, the difference appears. This comparison suggests that the effect of temperature should be considered especially when cell is operated in high current density conditions.

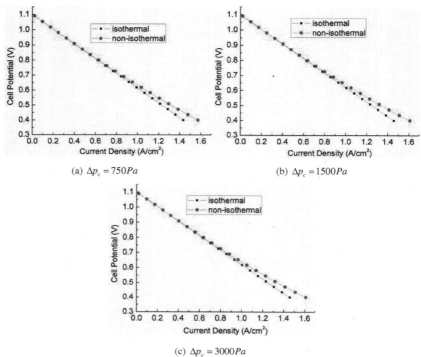

(a) $\Delta p_c = 750Pa$      (b) $\Delta p_c = 1500Pa$

(c) $\Delta p_c = 3000Pa$

Fig. 10 Cell performance comparison with non/isothermal assumption

### 4.4 Effects of fuel and gas flow directions

Fuel/gas flow settings may influence various aspects of cell performance, typically temperature distributions. Figure 11 shows temperature distributions for different flow direction settings, where the cell voltage is set at 0.4V and pressure loss is 750Pa. As one can see that, for co-flow condition, the high temperature region is close to the outlet (Figure 11 (a)); for counter-flow condition, the high temperature region seats in the middle of the cell (Figure 11 (b)); while for cross-flow condition, the high temperature region is at one of the corners (Figure 11 (c)).

The corresponding cell performances are shown in Figure 12 under above three flow settings. Clearly cell V-I curve is coincident with each other when current density is less than 0.6A/cm². Above 0.6A/cm², the cell with co-flow shows the best performance while the cell with counter-flow is the worst; the cell with cross-flow is in between. One reason is that beyond current density of 0.6A/cm², the effect of heat generated through electrochemical reaction becomes apparent. This result is in fact consistent with previous analysis in Figure 12 when isothermal and non-isothermal conditions are compared.

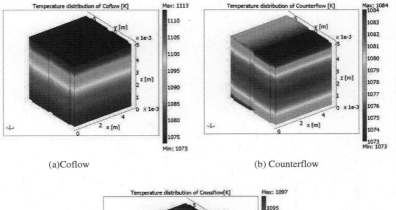

(a)Coflow                                    (b) Counterflow

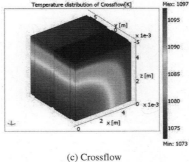

(c) Crossflow
Fig. 11 Temperature distributions under different flow settings

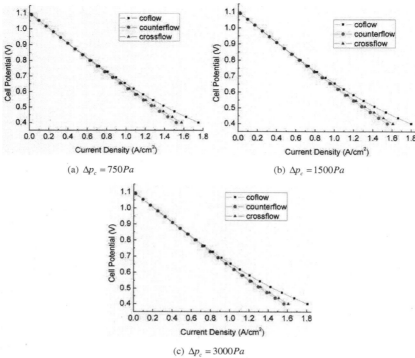

(a) $\Delta p_c = 750 Pa$

(b) $\Delta p_c = 1500 Pa$

(c) $\Delta p_c = 3000 Pa$

Fig. 12 Cell performance with different flow settings

## 5. CONCLUSIONS

A comprehensive 3D CFD model is developed for a SOFC with bi-electrode supported cell (BSC) design as a physical base. The model includes complicated transport phenomena of mass/heat transfer, charge (electron and ion) migration, and electrochemical reactions. The uniqueness of this study is that functionally graded porous electrode property is taken into account, which involves not only linear but also nonlinear porosity distributions in a general sense. The numerical code is validated using experimental data from open literature. Extensive numerical analysis is performed to elucidate both porous microstructure distribution and operating condition effects on cell performance from theoretical point of view. Results indicate that the cell with inverse parabolic porous electrode demonstrates promising performance; the optimal profile of inverse parabolic porosity distribution is dependent on operating conditions, typically pressure losses across electrodes. Increasing electrode thickness makes the V-I curve approach toward each other for the cells with linear and inverse parabolic porous electrodes. This trend is not apparent for the cell with parabolic porous electrode. In general, increasing electrolyte thickness leads to worse cell performance. Fuel/gas flow settings influence both cell temperature distribution and cell performance.

Nomenclature

| | |
|---|---|
| $A_V$ (m$^{-1}$) | Volumetric reaction surface area |
| $d_{pore}$ (m) | Particle diameter in porous electrode |
| $d_p$ (m) | Pore diameter in porous electrode |
| $D_{ij}$ (m$^2 s^{-1}$) | Binary mass diffusion coefficient of a mixture species $i$ and $j$ |
| $D_{ij}^{eff}$ (m$^2 s^{-1}$) | Effective diffusion coefficient |
| $D_{Kn,i}$ (m$^2 s^{-1}$) | Knudsen diffusion coefficient of species $i$ |
| $F$ (C mol$^{-1}$) | Faraday's constant: 96487 |
| $h$ | Enthalpy |
| $j_i$ (A m$^{-2}$) | Current density |
| $j_0$ (A m$^{-2}$) | Exchange current |
| $K$ (m$^2$) | Permeability of porous electrode |
| $M_{ij}$ | Mean molecular mass |
| $M$ (kg mol$^{-1}$) | Average molecular weight |
| $n_i$ | Number fraction |
| $N_t$ (m$^{-3}$) | Number density of all particles |
| $p$ (Pa) | Pressure |
| $p_i$ | Probability for $i$-phase particles to form percolated or globally continuous clusters |
| $R$ (J mol$^{-1}$K$^{-1}$) | Universal gas constant |
| $S$ (kg m$^{-3}s^{-1}$) | Mass source term |
| $S_h^e$ (W m$^{-3}$) | Heat generation due to electrochemical reaction |
| $S_h^j$ (W m$^{-3}$) | Heat generation due to joule heating |
| $S_i$ (kg m$^{-3}s^{-1}$) | Reaction source term for species $i$ |
| $T$ (K) | Temperature |
| $u$ (m $s^{-1}$) | Velocity |
| $V$ (V) | Potential |
| $x$ (m) | Direction parallel to electrolyte layer |
| $x_i$ | Mole fraction of species $i$ |
| $y$ (m) | Direction perpendicular to electrolyte layer |
| $Z$ | Total average coordination number |
| $Z_i$ | Total coordination for $i$-phase particles |

Greek symbols

| | |
|---|---|
| $\alpha$ | Electron transfer coefficient |
| $\mu$ (N s m$^{-2}$) | Viscosity |
| $\rho$ (kg m$^{-3}$) | Density |
| $\omega_i$ | Weight fraction of species $i$ |
| $\eta$ (V) | Over-potential |
| $\phi$ (V) | Exchange potential |
| $\gamma$ | Reaction order for oxidation or reduction |

| $k$(S m$^{-1}$) | Ionic conductivity |
|---|---|
| $\sigma$(S m$^{-1}$) | Electronic conductivity |
| $\tau$ | Tortuosity |
| $\varepsilon$ | Porosity |

Subscripts

| c | Cathode |
|---|---|
| a | Anode |
| $i$ | Species $i$ |
| ref | Reference |
| eq | Equilibrium |
| pol | Polarization |
| $el$ | Electronic |
| $io$ | Ionic |

Superscripts

| $eff$ | Effective |
|---|---|
| e | Electrochemical Reaction |
| j | Joule |

REFERENCES

[1]  S. Sunde, Journal of Electrochemical Society, 143 (3) pp.1123-1132, 1996.
[2]  C.W. Tanner, K.-Z. Fung, and A. V. Virkar, Journal of Electrochemical Society, Volume 144, pp.21-30, 1997.
[3]  E. Koep, C. Jin, M. Haluska, R. Das, R. Narayan, K. Sandhage, R. Snyder, M. Liu, Journal of Power Sources, Volume 161, pp.250-255, 2006.
[4]  R. E. Williford, P. Singh, Journal of Power Sources, Volume 128, pp.45-53, 2004.
[5]  B. R. Roy, N. M. Sammes, T. Suzuki, Y. Funahashi, M. Awano, Journal of Power Sources, Volume 188, pp.220-224, 2008.
[6]  H. Zhu, A. M. Colclasure, R. J. Kee, Y. Lin, S. A. Barnett, Journal of Power Sources, Volume 161, pp.413-419, 2006.
[7]  M.M. Hussain, X. Li, I. Dincer, Journal of Power Sources, Volume 161, pp.1012-1022, 2006.
[8]  Y. Shi, N. Cai, C. Li, Journal of Power Sources, Volume 164, pp.639-648, 2007.
[9]  J. H. Nam, D. H. Jeon, Electrochimica Acta, Volume 51, pp.3446-3460, 2006.
[10]  D. H. Jeon, J. H. Nam, and C.-J. Kim, Journal of The Electrochemical Society, 153 (2) A406-A417, 2006.
[11]  S. Ahmed, C. McPheeters, and R. Kumar, J. Electrochem. Soc., Volume 138, No. 9, September 1991
[12]  K. P. Recknagle, R. E. Williford, L. A. Chick, D. R. Rector, M. A. Khaleel, Journal of Power Sources, Volume 113, Issue 1, 1 January 2003, Pages 109-114
[13]  Y. Inui, N. Ito, T. Nakajima, A. Urata, Energy Conversion and Management, Volume 47, Issues 15-16, September 2006, Pages 2319-2328
[14]  D. Larrain, J. Van herle, F. Maréchal, D. Favrat, Journal of Power Sources, Volume 118, Issues 1-2, 25 May 2003, Pages 367-374
[15]  G. M. Goldin, H. Zhu, R. J. Kee, D. Bierschenk, S. A. Barnett, Journal of Power Sources, Volume 187, Issue 1, 1 February 2009, Pages 123-135.
[16]  M. F. Serincan, U. Pasaogullari, and N. M. Sammes, Journal of The Electrochemical Society, Volume 155 (11), 2008, B1117-B1127.
[17]  D. Larrain, J. Van herle, F. Maréchal, D. Favrat, Journal of Power Sources, Volume 118, Issues 1-2, 25 May 2003, Pages 367-374

[18]    T. L. Cable, S. W. Sofie, Journal of Power Sources, Volume 174, Issue 1, 2007, pp.221-227.
[19]    E.S. Greene, W.K.S. Chiu, and M.G. Medeiros, Journal of Power Sources, 161 (2006) 225-231.
[20]    M. Ni, M.K.H. Leung, and D.Y.C. Leung, Journal of Power Sources, 168 (2007) 369-378.
[21]    J. Deseure, Y. Bultel, L. Dessemond, E. Siebert, Electrochimica Acta, Vol. 50, pp 2037-2046, 2005.
[22]    D. H. Jeon, Electrochimica Acta, Volume 54, Issue 10, 2009, pp.2727-2736.
[23]    Alazmi, B, Vafai, K, Analysis of variants within the porous media transport models, Journal of Heat Transfer ASME, Volume 122, 2000, pp.303-326
[24]    M. Suzuki and T. Oshima, Powder Technology, Volume 35, Issue 2, 1983, pp.159-166
[25]    S. Sunde, Journal of Electroceram, Volume 5,2000, pp.153-182.
[26]    P. Costamagna, P. Costa, and V. Antonucci, Electrochimica Acta, Volume 43, pp.375-394, 1998.
[27]    Feng Zhao, Anil V. Virkar, Dependence of polarization in anode-supported solid oxide fuel cells on various cell parameters, Journal of Power Sources, Volume 141, 2005, pp.79-95

# FABRICATION AND PROPERTIES OF NANO-STRUCTURAL $Bi_2O_3$-$Y_2O_3$-$ZrO_2$ COMPOSITE

Jingde Zhang*, Kangning Sun, Jun Ouyang, Jian Gao, Han Liu, Tao An, Lijuan Xing

Key Laboratory for Liquid-Solid Structural Evolution and Processing of Materials (Ministry of Education), Shandong University
Jinan, Shandong, China

## ABSTRACT
$Bi_2O_3$ is an excellent sintering assistant and a good dopant for YSZ. In this work, a $Bi_2O_3$-$Y_2O_3$-$ZrO_2$ (Bi-YSZ) nano-structural composite was fabricated by a combined gel-casting and low temperature sintering process. The relationships between the amount of doped $Bi_2O_3$ and the conductivity, microstructure and phase ratios of the nanocomposite were investigated. The results showed an optimum volume ratio of 53% in solid content during the gel-casting process, and an optimum relative density of 97–99% in 3% $Bi_2O_3$ -YSZ sintered at 1000℃. The improvement in the conductivity of Bi-YSZ comes from the contributions of oxygen ion conductivity of $Bi_2O_3$ solved in YSZ, as well as the high conductivity of the δ-$Bi_2O_3$, which was formed in excessively doped samples. In addition, it was found that sintering processes significantly affect electrical performance, phase ratios and microstructures of Bi-YSZ.

## 1. INTRODUCTION
Due to its excellent material properties, such as superior oxygen ion conductivity, good mechanical properties, high oxidation resistance and corrosion resistance , stability for most electrode materials and so on,[1,2] Yttrium-Stabilized-Zirconium oxide (YSZ) solid electrolyte is the preferred material candidate for applications in oxygen sensors and solid oxide fuel cells (SOFC). However, there are also some drawbacks in YSZ, such as poor aging performance, low electrical conductivity, and tendency to crack. Consequently, more and more researchers have switched their focus from pure YSZ to the studies of molding and doping on this material[3-8].

As an electrolyte material, $Bi_2O_3$ has many advantages including very high electrical conductivity, low synthesizing temperature, and good sinterability. However, it also has some disadvantageous properties which have restricted its direct applications in SOFC.[9-18] Such properties include a very narrow temperature range for applications (730℃-850℃), poor mechanical properties and tendency of deoxidization.

Because of its low melting point, $Bi_2O_3$ can decrease the sintering temperature of YSZ when used as its sintering assistant. On the other hand, $Bi_2O_3$ can also improve the oxygen ion conductivity of YSZ significantly, due to its high oxygen ion conductivity at low temperature. In this work, a combined process of gel-casting and low temperature sintering was used to fabricate $Bi_2O_3$-doped YSZ nanocomposites for electrolyte applications. Moreover, the chemical composition, microstructure and electrical properties of the prepared nanocomposites were studied as functions of dopant amount and various process parameters.

## 2. EXPERIMENT METHODS
$Y_2O_3$ stabilized $ZrO_2$ (YSZ, 3 mol.% $Y_2O_3$) nano-powders (diameter ~70-100nm)were used as the base material. $Bi_2O_3$ nanopowders (diameter ~ 60-90nm) were used as additives. Acrylamide (AM) was used as the organic monomer while N,N'-methylene bis-acrylamide (MBAM) was used as

the cross-linker. In addition, $(NH_4)_2S_2O_8$ solution and N,N,N′,N′-tetramethyl ethylenediamine (TMED) were used as the polymerizing initiator and the catalyst, respectively[3].

The mixed powders of YSZ and $Bi_2O_3$ in different proportions and dispersants were added into monomer solutions, and ball-milled (in a QM-ISP planet high energy ball mill) for 20 hours to prepare the homogeneous suspension. The resulted suspension was vacuum de-aired, and added with initiator and catalyst, then poured into a die to form the green bodies. After drying and binder burnout, the green bodies were sintered in a high temperature box furnace (SRJX-8-13) under different conditions.

The microstructures of the sintered $Bi_2O_3$-doped YSZ were characterized by SEM (JEOL, JSM6610LA) and FE-SEM (Hitachi, S-4200). The crystalline structures were examined by X-ray diffraction (X'Pert Philips, PW1700). The density of the composites was measured by the Archimedes method. The electrical conductivity of the sample was measured by a four-probe DC technique.

## 3. RESULTS AND DISCUSSION

### 3.1 Gel-casting of $Bi_2O_3$-$Y_2O_3$-$ZrO_2$ materials

Fig.1-4 show the influence of pH, dispersant amount, solid content volume and milling time on the viscosity of the ceramic slurry. Fig.1 and 2 indicate that both the dispersant amount and the pH value affect the rheology of slurry significantly. The optimum amount of dispersant is around 1.6wt% and the optimum pH value is around 10. It can be seen from Fig. 3 that milling time also significantly affects the viscosity of the slurry, which decreases with the increase of milling time and levels off after 20h. Consequently, 20h was taken as the optimum milling time. Fig.4 indicates that, when the solid content is less than 50%, it has little influence on the viscosity. A critical solid content of 53% is observed, beyond which the viscosity increases very fast with solid content.

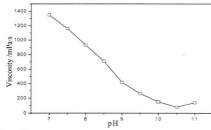

Fig.1 The effect of pH on viscosity of $Bi_2O_3$-YSZ slurry

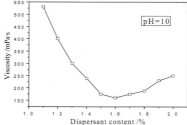

Fig.2 The relationship between dispersant content and viscosity

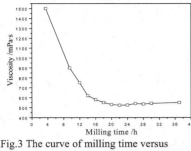

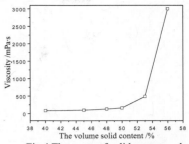

Fig.3 The curve of milling time versus viscosity

Fig.4 The curve of solid content volume versus viscosity

Fig.5 shows the effect of solid content volume on shrinkage. It is found that higher solid content of slurry leads to less shrinkage. Therefore, the optimum solid content for gel-casting was chosen at 53vol% by considering its effects on both viscosity and shrinkage.

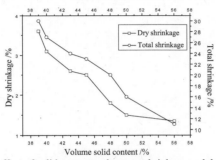

Fig.5 The effect of solid content volume on shrinkage and dry shrinkage

### 3.2 Sintering process and structures of Bi-YSZ

The sintering temperature in this work was set at 1000°C. This temperature point was chosen to be slightly higher than the melting point of $Bi_2O_3$ (825°C) [3] to provide liquid phase sintering of $Bi_2O_3$, therefore reducing the overall sintering temperature. The heating rate from room temperature to 600°C was set at 100°C/h and was changed to 50°C/h in the temperature range of 600°C-700°C, due multiple phase transitions of $Bi_2O_3$ in this temperature range (first from α- to γ-, and then to δ-$Bi_2O_3$)[4,5]. From 700°C to 1000°C the heating rate was changed back to 100°C/h. In order to investigate the influence of sintering process on the properties of $Bi_2O_3$-$Y_2O_3$-$ZrO_2$ composite materials, different sintering time and cooling methods were used and are shown schematically in Fig.6.

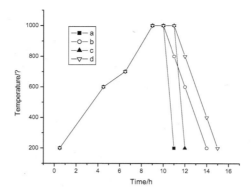

Fig.6 Sintering processes: (a) sintering for 1h, air cooling; (b) sintering for 1h, furnace cooling; (c) sintering for 2h, air cooling; and (d) sintering for 2h, furnace cooling

Fig.7 shows the relative density of 3mol%Bi-YSZ fabricated by different sintering processes. The relative densities of Bi-YSZ sintered at 1000□ were about 97–99%, and increased with the sintering time. The relative densities of furnace-cooled Bi-YSZ were only a little higher than those of air-cooled samples.

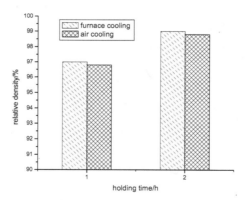

Fig.7 Relative density related to sintering time with different cooling methods

The cross-sectional SEM images of 3mol%Bi-YSZ fabricated by different sintering processes are shown in Fig.8. Grains with size of 0.1-0.7μm were obtained for the samples sintered at 1000°C. The samples sintered with longer sintering time (Fig.8 c & d) showed a slightly larger grain size than those in samples sintered with shorter sintering time (Fig.8 a & b). In addition, there was little difference in grain sizes between air-cooled samples (Fig.8 a & c) and furnace-cooled ones (Fig.8 b &

d). It indicates that the two cooling methods make little difference on affecting the resulting microstructures of Bi-YSZ. In Fig.8, it can also be seen that the grains were not uniform. Furthermore, Fig.9 shows the FE-ESM morphology of polished surfaces of the two samples sintered at 1000□ for 2h. It reveals that the grains have a broad size distribution, which include both nanocrystalline grains and clustered coarse grains (several micrometers).

Fig.10 shows the XRD patterns of Bi-YSZ with $Bi_2O_3$ contents of 3mol%, 5 mol% and 7mol%, sintered at 1000°C for 2h by air cooling. It shows that when the amount of $Bi_2O_3$ dopant was 3mol%, there was negligible amount of distinctive $Bi_2O_3$ phase in the resulting ceramics (Bi-YSZ). However, when the amount of $Bi_2O_3$ dopant increased to 5mol% and 7mol%, $Bi_2O_3$ phases can be distinctly identified in the resulting ceramics (separated phases of YSZ and $Bi_2O_3$). This means the solubility limit of $Bi_2O_3$ in YSZ is between 3mol% and 5mol%.

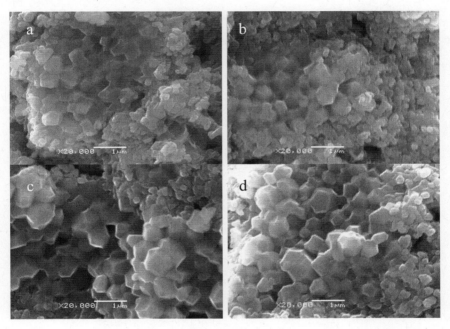

Fig.8 Cross-sectional SEM images of Bi-YSZ sintered at (a) sintering for 1h, air cooling, (b) sintering for 1h, furnace cooling, (c) sintering for 2h, air cooling, and (d) sintering for 2h, furnace cooling

Fig.9 FE-ESM morphology of Bi-YSZ sintered at 1000°C for 2h followed by (a) air cooling, and (b) furnace cooling.

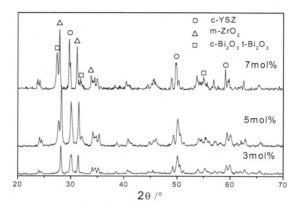

Fig.10 XRD patterns of Bi-YSZ with different $Bi_2O_3$ contents

## 3.3 Electrical properties of Bi-YSZ

Fig.11 shows the electrical conductivity of Bi-YSZ sintered at 1000°C for 1h followed by furnace cooling, with $Bi_2O_3$ content of 3mol%, 5 mol% and 7mol%. It is found that the electrical conductivity increases with $Bi_2O_3$ content at the same temperature. A brief description of the underlying mechanism is provided as following: the solid solution of $Bi_2O_3$ in YSZ can produce oxygen vacancies, which results in higher conductivity of Bi-YSZ, while the doping of excessive $Bi_2O_3$ formed δ-$Bi_2O_3$ (shown in Fig.10), which has high conductivity and hence further improves the conductivity of Bi-YSZ.

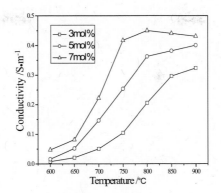

Fig.11 Electrical conductivity of Bi-YSZ sintered at 1000°C for 1h followed by furnace cooling

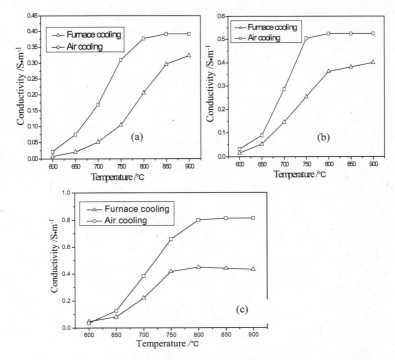

Fig.12 Conductivity-temperature curve of a) 3mol%, b) 5mol% and c) 7mol%
$Bi_2O_3$ doped YSZ in different cooling method

The influence of cooling methods on the electrical conductivity is schematically shown in Fig.12. It can be seen that, under the same doping and sintering conditions, air-cooled Bi-YSZ composites have higher electrical conductivities that those of furnace-cooled samples. This is due to the "quenching" effect induced by the air-cooling process. After air-cooling, metastable high temperature phase structures, which have higher fraction of Bi$_2$O$_3$-YSZ solid solutions, remain in the composite materials, and therefore showed higher electrical conductivity.

Fig.13 shows the influence of sintering time on the electrical conductivity. As shown in Fig.13 a & b, the electrical conductivities of 3mol%Bi-YSZ and 5mol%Bi-YSZ were strong functions of the sintering time – samples sintered longer showed higher electrical conductivities at the same temperature. However, Fig. 13 c showed that, for the 7mol%Bi-YSZ, the influence of sintering time on its electrical conductivity is not significant. The above phenomena can be explained by the following mechanism: longer sintering time results in larger grains, and larger grains correspond to lower activation energy of the mobility of oxygen ions, which leads to a decreased electrical conductivity[6]. However, in the 7mol% samples, the overall electrical conductivity was dominated by the contribution from the electrical conductivity of the excessive Bi$_2$O$_3$, which is the same in the two samples with different sintering time, therefore the two samples showed comparable conductivities.

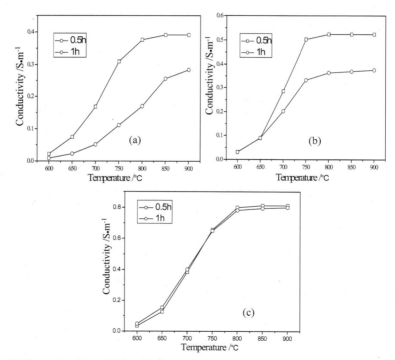

Fig.13 The conductivity of a) 3mol%, b) 5mol% and c) 7mol% Bi-YSZ by different sintering time.

# 4. CONCLUSIONS

A $Bi_2O_3$-$Y_2O_3$-$ZrO_2$ composite was fabricated by a combined process of gel-casting and low temperature sintering. An optimized solid content of the green bodies in terms of adequate viscosity and low shrinkage, was discovered at 53vol%. This solid content was used for the gel-casting process.

3mol% $Bi_2O_3$-YSZ sintered at only 1000°C showed an almost fully-dense ceramic structure. The relative densities were measured at about 97–99%. The grains of the ceramics showed a broad size distribution, with many nanocrystalline grains and some clustered coarse grains. The above results indicate that $Bi_2O_3$ significantly decreases the sintering temperature of YSZ as a sintering assistant. Additionally, the solubility limit of $Bi_2O_3$ in YSZ is found to be between 3mol% and 5mol%. Furthermore, doping of $Bi_2O_3$ significantly improves the electrical conductivity of YSZ, which has contributions from the oxygen ion conductivity of solid solution of $Bi_2O_3$ in YSZ, as well as electrical conductivity from the segregated δ-$Bi_2O_3$ phase in heavily doped Bi-YSZ samples.

## ACKNOWLEDGEMENTS
This paper was supported by the Nature Science Foundation of the People's Republic of China (No.50942024, 50872072) and Excellent Young and Middle-aged Scientist Foundation of Shandong Province (No. 2007BS04048).

## REFERENCES
[1] Ch. Laberty-Robert, F. Ansart, C. Deloget, et al. Powder synthesis of nanocrystalline $ZrO_2$-8%$Y_2O_3$ via a polymerization route[J]. Mater. Res. Bull. 2001, 36(12):2083~2101.
[2] Y. Sakaki, Y. Takeda, A. Kato, et al.$Ln_{1-x}Sr_xMnO_3$ (Ln=Pr, Nd, Sm and Gd) as the cathode material for solid oxide fuel cells[J]. Solid State Ionics , 1999, 118 (3-4): 187-194.
[3] Qiang Qiang Tan, Min Gao, Zhong Tai Zhang and Zi Long Tang. Polymerizing mechanism and technical factors optimization of nanometer tetragonal polycrystalline zirconia slurries for the aqueous-gel-tape-casting process. Materials Science and Engineering A, 2004, 382(1-2) :1-7.
[4] J.W. Medernach, R.L. Snyder. Powder diffraction patterns and structures of the bismuth oxides[J]. J. Am. Ceram. Soc., 1978,61:494-497.
[5] Arora P, White R E, Doyle M. Capacity Fade Mechanisms and Side Reactions in Lithium-Ion Batteries[J]. J. Electrochem. Soc., 1998,145(10):3647-3667.
[6] Lai Wei, Haile Sossina M. Impedance Spectroscopy as a Tool for Chemical and Electrochemical Analysis of Mixed Conductors: A Case Study of Ceria[J]. Journal of the American Ceramic Society, 2005,88(11):2979-2997.
[7] Seung-Goo Kima, Sung Pil Yoonb, Suk Woo Nam,et al. Fabrication and characterization of a YSZ/YDC composite electrolyte by a sol–gel coating method. Journal of Power Sources. 110 (2002) 222–228.
[8] Min-Fang Han, Hui-Yan Yin, Wen-Ting Miao, Su Zhou. Fabrication and properties of anode-supported solid oxide fuel cell. Solid State Ionics. 179 (2008) 1545–1548.
[9] Sato T, Ohtaki S, Shimada M. Transformation of yittria patially stabilized zirconia by temperature annealing in air[J]. J Mater Sci, 1985,20:1466.
[10] J.W . Medernach and R.L. Snyder, "Powder diffraction patterns and structures of the bismuth oxides," J.Am. Ceram. Soc., 61, 1978:494-497.
[11] Arora P, White R E, Doyle M. Capacity Fade Mechanisms and Side Reactions in Lithium-Ion Batteries[J]. J. Electrochem. Soc., 1998,145(10):3647-3667.
[12] Sammes, N.M. Bismuth based oxide electrolytes-structure and ionic conductivity[J]. Journal of the European ceramic society, 1999, 19: 1801-1826.

[13]Watanabe,A,Kikuchi,T.,Cubic-Hexagonal Transformation of Yttria- Stabilized δ-Bismuth Sesquixide $Bi_{2-2x}Y_2XO_3$(x=0.215-0.235)[J]. Solid State Ionics, 1986, 21:287-291.

[14]Takahashi, T., Iwahara, H., Oxide ion conductors based on bismuth sesquixide[J]. Materials Research Bulletin, 1987, 13: 1447-1453.

[15]Conflant, P., Boivin, J.C., and Thomas, D., Le diaggramme des phases solides du systeÂ me Bi2O3-CaO [J]. Journal of Solid State Chemistry, 1986,18:133-140.

[16]Sillen, L.G.,Aurivillius, B. Oxide phases with adefect oxygen lattice[J]. Zeitschrift fuÊr Kristallographie, 1939,101:483-495.

[17]Takahashi, T., Iwahara, H., and Nagai, Y., High oxide ion conduction in sintered bismuth oxide containingstrontium oxide, calcium oxide, or lanthanum oxide[J]. Journal of Applied Electrochemistry, 1972,2:97-104.

[18]Sammes, N.M., Bismuth based oxide electrolytes-structure and ionic conductivity [J] . Journal of the European cetamic society, 1999, 19:1801-1826 .

*Corresponding author.  E-mail: zhangjingde@sdu.edu.cn (J.Zhang).

# Author Index